AF813503

Francesco Aldo Costabile

Modern Umbral Calculus

De Gruyter Studies in Mathematics

Edited by
Carsten Carstensen, Berlin, Germany
Gavril Farkas, Berlin, Germany
Nicola Fusco, Napoli, Italy
Fritz Gesztesy, Waco, Texas, USA
Niels Jacob, Swansea, United Kingdom
Zenghu Li, Beijing, China
Karl-Hermann Neeb, Erlangen, Germany

Volume 72

Francesco Aldo Costabile

Modern Umbral Calculus

An Elementary Introduction with Applications to Linear Interpolation and Operator Approximation Theory

DE GRUYTER

Mathematics Subject Classification 2010
Primary: 05A40; Secondary: 41A10

Author
Prof. Dr. Francesco Aldo Costabile
Università della Calabria
Dipartimento di Matematica
e Informatica
Ponte Pietro Bucci 30B
87036 Cosenza
Italy
francesco.costabile@unical.it

ISBN 978-3-11-064996-3
e-ISBN (PDF) 978-3-11-065292-5
e-ISBN (EPUB) 978-3-11-065009-9
ISSN 0179-0986

Library of Congress Control Number: 2019937570

Bibliographic information published by the Deutsche Nationalbibliothek
The Deutsche Nationalbibliothek lists this publication in the Deutsche Nationalbibliografie; detailed bibliographic data are available on the Internet at http://dnb.dnb.de.

© 2019 Walter de Gruyter GmbH, Berlin/Boston
Typesetting: VTeX UAB, Lithuania
Printing and binding: CPI books GmbH, Leck

www.degruyter.com

To my family

Preface

In the title of this book is the word "Modern", but—really—we should use the expression "Modern Classical". In fact, within the past years, different forms of umbral calculi have begun to be studied.

The Umbral Calculus was described for the first time by John Blissard in the 1850's ([23]) in a form that we call "Classical". After a short phase of success, the Umbral Calculus was largely rejected by the mathematics community due to the "lack of rigor" ([166]).

In the late 1960s the theory, worked out by Gian Carlo Rota and his co-workers, gave a completely rigorous formulation to the Umbral Calculus, which greatly rehabilitated it. The work [167] and the book [166] give an extensive and lucid presentation of the Umbral Calculus, whereas a shorter introduction can be found in [93]. The "Modern Classical" Umbral Calculus is now the systematic study of Sheffer polynomial sequences, including Binomial and Appell sequences. In fact, the Umbral Calculus, in Rota's acceptation, allows an algebraic treatment of classical polynomials and numbers beginning from generating functions, recursive and reciprocity formulas, expansion theorems et cetera, depending on the choice of the formal power series as the "Umbra" (Latin for Shadow) of linear functionals and polynomials. Therefore, the Umbral Calculus is a mix of linear algebra, theory of formal power series, and classical analysis.

The Modern Umbral Calculus has been approached from different points of view. For example, by formal power series ([165, 166, 179]), algebraic ([170, 193]), or operator theoretic ([146, 168]). Each of these approaches has been followed by many authors in different applications (see, for example, [5, 10, 38, 78, 81–83, 104, 146]).

In recent times, it has been observed that there is an isomorphism between the Riordan matrices (see, for example, [5, 104, 177]) and the Sheffer polynomials ([179]) (and hence also the Appell polynomials [17] and Binomial sequences [93]). At the same time, the possibility to define the Sheffer polynomials through determinantal forms has been proved (see [60, 62–66, 213, 215]).

Based on these dernier papers, in this book there is an attempt to present a theory of Modern Umbral Calculus in one variable, that is, known and also unknown results, using essentially elementary matrix calculus: lower triangular, infinite matrices, Hessemberg, Toeplitz, Riordan-type matrix, determinant and Cramer's rule, recurrence relations, and few more. Hence, this book is not a complete and updated survey, but a new approach to the classic umbral calculus. In truth, the use of matrices in the theory of umbral calculus goes back to Vein's papers ([197, 198]). In particular, in ([198]) it is written: "The referee printed out that this work is an explicit matrix version of umbral calculus as presented in Rota et al. ([167, 168, 170])". This work shows that Vein's procedures are really different from ours. Our procedures are simpler and for a

https://doi.org/10.1515/9783110652925-201

wider audience. In Section 1.2 of Chapter 1, Vein's approach will be sketched and the differences will be clarified.

The motivation of our choice is to target the largest number of readers: from undergraduate students to young researchers, even those in disciplines other than mathematics. We also stress the importance of umbral calculus in the training of young students in mathematics.

The modern umbral calculus has more applications and in various disciplines: probability theory (for example, [38, 74, 78, 82, 83, 169]), number theory (for example, [87, 96]), linear recurrence (for example, [146]), et cetera. In the sequel, we consider the applications to general linear interpolation (for example, [61, 62, 67, 128, 200]) and operators approximation theory (for example, [10, 112, 152, 188, 189]). Moreover, we point out that a sufficiently comprehensive bibliography up to 2000 is in [79].

In the Chapter 1 of Part I, there is a more detailed presentation of the contents of the book.

In closing this preface, I would like to quote H. J. Stetter ([187]) and G. Walz ([203]): "*I ventured to write this book in English because it will be more easily read in poor English than in good Italian by* 90 % *of my intended readers.*"

Rende, May 2018 *Francesco Aldo Costabile*

Acknowledgment

I would like to express my gratitude to M. Altomare, M. I. Gualtieri, and A. Napoli to have composed the text in Latex, and subsequent to their reading of the manuscript, proposed some interesting changes to the same.

https://doi.org/10.1515/9783110652925-202

Contents

Preface —— VII

Acknowledgment —— IX

Acronyms —— XV

Part I: **Introduction**

1 Preliminaries and notations —— 3
1.1 Polynomial sequences —— **3**
1.2 About polynomial sequences represented by lower triangular, nonsingular infinite matrices —— **4**
1.3 The plan of the book —— **5**

2 Particular matrices and their connections with formal power series —— 7
2.1 Binomial-type matrix —— **7**
2.2 Appell-type matrix —— **9**
2.3 Sheffer-type matrix —— **11**
2.4 Lidstone-type matrix —— **15**
2.5 Production matrix —— **18**
2.6 Summary —— **19**

Part II: **Polynomial sequences of binomial type**

3 Binomial polynomial sequences —— 23
3.1 Introduction —— **23**
3.2 Definition and characterizations —— **23**
3.3 Recurrence relations —— **27**
3.4 Determinantal forms —— **31**
3.5 Relationship with δ-functionals —— **34**
3.6 Relationship with δ-operators —— **35**
3.7 Connection constants —— **38**
3.8 Summary —— **40**

4 Applications to linear interpolation and operators approximation theory —— 41
4.1 Linear interpolation —— **41**
4.2 Approximation operators of binomial type —— **42**

4.3 Summary — 45

5 Examples — 47
5.1 The sequence $\{x^n\}_{n\in\mathbb{N}}$ — 47
5.2 Lower factorial and exponential polynomials — 48
5.3 Abel polynomial sequences — 55
5.4 Binomial Laguerre polynomials — 62
5.5 Central factorial polynomials — 67

Part III: Appell polynomial sequences

6 Appell polynomial sequences — 77
6.1 Introduction — 77
6.2 Definition and characterizations — 78
6.3 Recurrence relations — 81
6.4 Determinantal forms — 85
6.5 General properties — 88
6.6 Localization of zeros — 93
6.7 Operational matrices — 95
6.8 Relationship with linear functionals — 96
6.9 Generalization of Appell polynomials — 98
6.10 Summary — 99

7 Application to linear interpolation and approximation theory — 101
7.1 Appell interpolation — 101
7.2 Operators approximation theory — 103

8 Examples — 107
8.1 Bernoulli polynomials — 107
8.2 Euler polynomials — 123
8.3 Hermite polynomials — 134
8.4 Appell–Laguerre polynomials — 141

Part IV: Sheffer polynomial sequences

9 Sheffer polynomial sequence — 149
9.1 Introduction — 149
9.2 Definition and first characterizations — 149
9.3 Recurrence relations and determinantal forms — 153
9.4 Differential equations — 158
9.5 General properties — 162

9.6 Relationship with δ-operators and functionals — 163
9.7 Summary — 165

10 Applications to linear interpolation and operators approximation theory — 167
10.1 Linear interpolation — 167
10.2 Operators approximation theory — 171

11 Examples — 175
11.1 The generalized Bernoulli polynomials of the second kind — 175
11.2 Generalized Laguerre polynomials — 184
11.3 Generalized Boole polynomials — 190
11.4 Poisson–Charlier polynomials — 195

Part V: Lidstone polynomial sequences

12 Lidstone-type polynomial sequences — 203
12.1 Introduction — 203
12.2 Odd Lidstone-type polynomial sequences — 203
12.3 Recurrence relation, determinantal form — 207
12.4 Even Lidstone-type polynomial sequences — 210
12.5 Recurrence relation, determinantal form — 213
12.6 Summary — 215

13 Application to linear interpolation and operators approximation theory — 217
13.1 Odd Lidstone-type linear interpolation problem — 217
13.2 Odd Lidstone-type generalized Szasz operators — 218
13.3 Even Lidstone-type interpolation problem — 219
13.4 Even Lidstone-type generalized Szasz operators — 220
13.5 Summary — 221

14 Examples — 223
14.1 First type Lidstone polynomials — 223
14.2 Odd Lidstone–Euler polynomials — 229
14.3 Second type or complementary Lidstone polynomials — 233
14.4 Even Lidstone–Euler polynomials — 238

Bibliography — 245

Index — 255

Acronyms

$\mathbb{K}$	field of characteristic 0 (tipically $\mathbb{R}$ or $\mathbb{C}$)
$(b_i)_i$	numerical sequence of elements of $\mathbb{K}$ with $b_0 = 0$ and $b_1 \neq 0$
$X_\infty \equiv X = (x_0, \ldots, x_n, \ldots)^T$	infinite column vector of entries x_i, $i = 0, 1, \ldots$
$X_n = (x_0, \ldots, x^n)^T$	column vector of entries $x^i \in \mathbb{K}$, $i = 0, 1, \ldots, n$
$\mathcal{P}$	set of polynomials with coefficients in $\mathbb{K}$
$\mathcal{P}_n$	set of polynomials of coefficients in $\mathbb{K}$ with *degree* $\leq n$
$\{p_n\}_n$	sequence of polynomials with degree $p_n = n$
$p_n(x)$	polynomial of *degree* n in the variable x
$f(x) := \sum_{i=1}^{\infty} b_i \frac{x^i}{i!}$	generating function of $(b_i)_i$, formal power series, δ-series
$(\widehat{b}_i)_i$	conjugate sequence of $(b_i)_i$
$\overline{f}(x) := \sum_{i=1}^{\infty} \widehat{b}_i \frac{x^i}{i!}$	compositional inverse of $f(x)$, formal power series, δ-series
$P := (p_{i,j})$	infinite lower triangular Binomial-type matrix
$\widehat{P} := (\widehat{p}_{i,j})$	conjugate matrix of P
$(a_i)_i$	numerical sequence of elements of $\mathbb{K}$ with $a_0 \neq 0$
$g(x) := \sum_{i=0}^{\infty} a_i \frac{x^i}{i!}$	generating function of (a_i), invertible power series
$(\widehat{a}_i)_i$	conjugate sequence of $(a_i)_i$
$\frac{1}{g(x)} = \sum_{i=0}^{\infty} \widehat{a}_i \frac{x^i}{i!}$	generating function of $(\widehat{a}_i)_i$, reciprical power series of $g(x)$
$A := (a_{i,j})$	infinite lower triangular Appell-type matrix
$\widehat{A} := (\widehat{a}_{i,j})$	conjugate matrix of A
$A_n := (a_{i,j})$	matrix of order n
$S := (s_{i,j})$	infinite lower triangular Sheffer-type matrix
$S := (g(x), f(x))$	exponential Riordan matrix
$\widehat{S} := (\widehat{s}_{i,j})$	conjugate matrix of S
$O := (o_{i,j})$	infinite lower triangular odd Lidstone-type matrix
$E := (e_{i,j})$	infinite lower triangular even Lidstone-type matrix
$(x)_n$	lower factorial polynomial of order n
$S(n,k)$	Stirling number of second kind
$s(n,k)$	Stirling number of first kind
$\phi_n(x)$	exponential polynomial of degree n
$\{L_n^{(\alpha)}\}_n$	Laguerre polynomial sequence
$\{p_n := n! L_n^{(-1)}\}_n$	binomial Laguerre polynomial sequence
$x^{(n)}$	central factorial polynomial of degree n
$\{a_n\}_n$	Appell sequence
$\mathcal{D}$	derivation matrix
$\mathcal{I}$	integration matrix
$P_n(f)(x)$	polynomial operator of order n
$\{B_n\}_n$	Bernoulli polynomial sequence

https://doi.org/10.1515/9783110652925-203

$\{\widehat{B}_n\}_n$	conjugate Bernoulli polynomial sequence
B_n	Bernoulli number
$\{E_n\}_n$	Euler polynomial sequence
$\{\widehat{E}_n\}_n$	conjugate Euler polynomial sequence
$\{H_n\}_n$	Hermite polynomial sequence
$\{s_n\}_n$	Sheffer sequence
$\{B^{\mathrm{II}}_{n,h}\}_n$	Bernoulli polynomial sequence of second kind
$\{\mathcal{B}_n\}_n$	generalized Boole polynomial sequence
$\{c_n\}_n$	Poisson–Charlier polynomial sequence
$\mathrm{OLP}(x)$	set of odd Lidstone-type polynomial sequence
$\mathrm{ELP}(x)$	set of even Lidstone-type polynomial sequence
$\{\Lambda_n\}_n$	Lidstone polynomial sequence of first type
$\{\mathcal{E}_n\}_n$	odd Lidstone–Euler polynomial sequence
$\{v_n\}_n$	Lidstone polynomial sequence of second type
$\{\mathcal{S}_n\}_n$	even Lidstone–Euler polynomial sequence
$\mathbb{N}$	set of non negative integer

Part I: **Introduction**

1 Preliminaries and notations

1.1 Polynomial sequences

This monograph focuses mainly on certain polynomial sequences. We should begin with some basic notations and definitions.

Let $\mathbb{K}$ be a field of characteristic 0 (typically $\mathbb{R}$ or $\mathbb{C}$). We denote by $(a_i)_i$ a numerical sequence of elements of $\mathbb{K}$. Often, we will use a single letter to indicate this sequence, that is, we will write $a := (a_i)_i$.

With capital letters, we denote matrices and vectors; the elements will be indicated with lowercase letters with two indices, respectively one, that is,

$$A := (a_{i,j})_{i,j\geq 0}, \quad X := [x_0, x_1, \dots, x_n, \dots]^T,$$

which are, respectively, the matrix A with elements $a_{i,j}$, $i, j = 0, 1, \dots$, and the vector X with components x_i, $i = 0, 1, \dots$.

For indicating the order of a matrix or a vector, we will use a subindex. That is, A_n and X_n are, respectively, a matrix and a vector of order n. We consider also infinite matrices and vectors, and we will indicate them with capital letters, such as A and X.

Moreover, we denote by $\mathcal{P}$ the set of polynomials in one variable and with $\mathcal{P}_n$, the set of polynomials of degree $\leq n \in \mathbb{N}$. A single polynomial will be indicated with a lowercase letter, that is, $p_n(x)$ is a polynomial in the variable x of degree exactly $n \in \mathbb{N}$.

With polynomial sequence, we indicate a set of polynomials denoted with $\{p_n\}_n$ that satisfies

$$\begin{cases} p_0(x) \neq 0, \\ \text{degree} \quad p_n(x) = n, \quad n \in \mathbb{N}. \end{cases} \tag{1.1}$$

For a polynomial $p_n(x)$, we use its decomposition on the monomial power basis:

$$p_n(x) = \sum_{k=0}^{n} t_{n,k} x^k, \quad t_{n,k} \in \mathbb{K}, \quad k = 0, \dots, n, \quad n \in \mathbb{N}. \tag{1.2}$$

By setting

$$\begin{aligned} T_n &:= (t_{i,k}), \quad i = 0, 1, \dots, n, \quad k = 0, 1, \dots, i, \\ X_n(x) &:= [1, x, \dots, x^n]^T, \end{aligned} \tag{1.3}$$

we have that $T_n X_n(x)$ gives the set of polynomials $\{p_0(x), \dots, p_n(x)\}$, that is,

$$P_n(x) = T_n X_n(x), \tag{1.4}$$

where $P_n(x)$ denotes the vector column $P_n(x) = [p_0(x), \dots, p_n(x)]^T$.

https://doi.org/10.1515/9783110652925-001

We note that T_n is a nonsingular, lower triangular matrix.

Likewise, for a polynomial sequence $\{p_n\}_n$, we can write $TX(x)$, where T is a nonsingular, lower triangular, infinite matrix, and $X(x)$ is the infinite column vector of classic monomials. Hence, for the infinite set of polynomials $\{p_0(x), \ldots, p_n(x), \ldots\}$, we get

$$X(x) := [1, x, \ldots, x^n, \ldots]^T,$$

$$T := \begin{pmatrix} t_{0,0} & & & & \\ t_{1,0} & t_{1,1} & & & \\ \vdots & & \ddots & & \\ t_{n,0} & t_{n,1} & \cdots & t_{n,n} & \\ \vdots & & \cdots & & \ddots \end{pmatrix},$$

$$P(x) := [p_0(x), \ldots, p_n(x), \ldots]^T,$$

and

$$P(x) = TX(x). \tag{1.5}$$

Formulas (1.4) and (1.5) are called *matricial forms* of a finite set or of a sequence of polynomials, respectively.

1.2 About polynomial sequences represented by lower triangular, nonsingular infinite matrices

As we have seen in the previous section, a polynomial sequence (p. s.) $\{p_n\}_n$ can be represented by the matricial form

$$P(x) = TX, \tag{1.6}$$

where

$$P(x) = [p_0(x), p_1(x), \ldots, p_n(x), \ldots]^T, \quad X = [1, x, x^2, \ldots, x^n, \ldots]^T,$$

and $T = (t_{i,j})_{i,j\geq 0}$ with $t_{i,j} = 0$ for $j > i$.

The concept of representing polynomial sequences by lower triangular, infinite matrices is not new and goes back to G. Polya (1928) and L. Schur [176]. In fact, Polya [151] gave a solution of the Cauchy–Bellman functional equation for matrices

$$M(x)M(y) = M(x + y) \tag{1.7}$$

in the form

$$\begin{pmatrix} 1 & 0 & \cdots & \cdots & \cdots \\ x & 1 & 0 & \cdots & \\ x^2 & 2x & 1 & \ddots & \\ x^3 & 3x^2 & 3x & 1 & \ddots \\ \vdots & & & & \ddots \end{pmatrix}. \tag{1.8}$$

Then, Vein [197] observed that the $(n+1)$th row of (1.8) contains the terms of the polynomial expansion of $(1+x)^n$, and the elements in column $(n+1)$ are the terms in the infinite series expansion of $(1-x)^{-n-1}$. Moreover, Vein for (1.7) expected a relation of the form

$$M(x) = e^{xQ}, \tag{1.9}$$

where Q is a triangular infinite matrix with constant elements. He proved relation (1.9) in [198] and observed that

$$Q = M'(0). \tag{1.10}$$

From this relation, Vein [198] proved some identities among triangular matrices and inverse relations. Thereafter, he determined two sets of triangular matrices. The elements of one set are related to the terms of Laguerre, Hermite, Bernoulli, Euler, and Bessel polynomials, whereas the elements of the other set consist of Stirling numbers of both kinds, the two-parameter Eulerian numbers by Touchard. It is then shown that these matrices are related by a number of identities. Some well-known and less well-known pairs of inverse scalar relations arising in combinatorial analysis are shown to be derivable from simple and obviously inverse pairs of matrix relations. Vein [198] said: "The referee has pointed out that this work is an explicit matrix version of the Umbral Calculus as presented by Rota et al. ([166, 168, 170])".

Our aims in the following are the same, that is, to find well-known results on the umbral calculus, but—also—some new identities and proprieties of Sheffer polynomial sequences, by elementary matrix calculus. Our approach, however, is very different from that of Vein. In fact, our starting point is the matrix form (1.4)–(1.5) and not (1.7)–(1.8)–(1.9)–(1.10), which indeed will be never considered.

The representation of a p. s. by a matrix can be found also in the Schur's paper on Faber polynomials [176], whereas in recent years more authors approached the umbral calculus by algebraic methods (we underline the papers [60, 62, 64, 66, 213–215]).

1.3 The plan of the book

In this Section, we will illustrate the content of the book, providing some motivational reasons regarding the pedagogical choices. Our preference is to follow a chronological (historical) order in the construction of three classes of polynomial sequences: binomial type, Appell, and Sheffer A-type 0.

However, whereas it is clear that through the papers of Appell [17] and Sheffer [179] we can precisely date the respective classes of polynomials (that take their name from the authors), we cannot do the same for the polynomials of the binomial type.

To the knowledge of the author, the p. s. of binomial type goes back to Bell [21], but also to Aitken's paper [11]. These works are on very different subjects: the first one is on combinatorial type and the other is on linear, general interpolation problem.

As will be clear in the following, both the classes, binomial type and Appell, are subclasses of the Sheffer A type-0 polynomial sequences. Furthermore, the binomial type are very close to the Sheffer classes.

In the sequel, we will deal with: binomial-type polynomial sequences (Part II), Appell polynomial sequences (Part III), Sheffer polynomial sequences of A-type 0 (Part IV).

Part V is dedicated to the so-called Lidstone-type polynomials. Lidstone [128] generalized Aitken's paper [11]. He gave a new basis polynomial helpful in the general interpolation problem related to the differential operator D^2. This class of polynomials has not been considered by Rota and other authors; therefore, it has not been included in the classic modern umbral calculus. But the deep correlation with the Appell polynomials motivated our choice to consider them here.

In Chapter 2 of Part I, we will consider particular classes of matrices and formal power series. Initially, the reader may not appreciate it, as he will not find the rationale for the algorithms of construction matrices. However, thereafter, he will be able to satisfy his curiosity, by discovering for himself the different equivalent reasons that gave rise to the construction of the algorithms.

Parts two, three, four, and five are divided into three chapters: theory, application to general linear interpolation and operator theory approximation, and examples.

Throughout the book, known and new identities and properties, such as recurrence relations and determinant forms, are given.

2 Particular matrices and their connections with formal power series

Abstract: In order to introduce special sequences of polynomials in matricial form, we define some particular classes of infinite, lower triangular matrices. They will be called binomial-, Appell-, Sheffer- and Lidstone-type matrices. Exponential Riordan and Production matrices will be considered too.

2.1 Binomial-type matrix

Let $\mathbb{K}$ be a field of characteristic 0 (typically $\mathbb{R}$ or $\mathbb{C}$), and let $b := (b_i)_i$ be a sequence of elements of $\mathbb{K}$ such that $b_0 = 0$, $b_1 \neq 0$. We will define a lower triangular, infinite matrix B ([199, 201], [217, Def. 1]) as follows.

Algorithm 2.1.1 (Construction of the matrix B). For any $N \in \mathbb{N}$, from the sequence $(b_i)_i$, we can define the following elements of $\mathbb{K}$:

$$\begin{aligned}
&\text{for } l = 0, 1, \ldots, N;\\
&\quad b_{l,0} = \delta_{l,0};\\
&\quad b_{l,1} = b_l;\\
&\quad\quad \text{for } k = 2, \ldots, l;\\
&\quad\quad\quad b_{l,k} = \frac{1}{k}\sum_{i=1}^{l-k+1}\binom{l}{i} b_{i,1} b_{l-i,k-1};\\
&\quad\quad \text{for } k = l+1, \ldots, N;\\
&\quad\quad\quad b_{l,k} = 0,
\end{aligned} \tag{2.1}$$

where $\delta_{i,j}$ is the Kronecker delta.

Then, we set $B_N := (b_{l,k})_{l,k}$, $l = 0, \ldots, N$, $k = 0, \ldots, l$.

The matrix B_N is called *binomial-type matrix of order N* .

For $N \to +\infty$, we get an infinite, lower triangular matrix of binomial type, and we denote it with B_∞ or, for simplicity, B. We note that B_N is the principal submatrix of B, of order N.

Remark 2.1. We note that

$$b_{i,i} = b_1^i, \quad \forall i = 1, 2, \ldots .$$

https://doi.org/10.1515/9783110652925-002

Proposition 2.1. *Let $f(x)$ be the formal power series associated with the sequence b, that is,*

$$f(x) = \sum_{i=0}^{\infty} b_i \frac{x^i}{i!}, \tag{2.2}$$

resulting in

$$\frac{1}{k!}(f(x))^k = \sum_{i=0}^{\infty} b_{i,k} \frac{x^i}{i!} \quad k = 0, 1, \dots. \tag{2.3}$$

Proof. We note that the power of a series is the repeated product of the series, and then the result follows by usual Cauchy product [150] and (2.1). □

The formal power series $f(x)$ is called the *generating function* of sequence b. We note that the elements of the kth column (being the first column indexed with 0) of the matrix B are the coefficients of the power series $\frac{1}{k!}(f(x))^k$ with $f(x)$ given in (2.2).

Remark 2.2. We note that in [78], the result is

$$b_{n,k} = \frac{1}{k!} \sum \binom{n}{\nu_1, \dots, \nu_k} b_{\nu_1} \cdots b_{\nu_k}, \tag{2.4}$$

in which the summation is over all k-tuples $(\nu_1, \dots, \nu_k)$, with $\nu_i \geq 1$ and $\sum_i \nu_i = n$.

Remark 2.3. If $b_i \geq 0$, $i = 0, 1, \dots$, the result is $b_{n,k} \geq 0$, $\forall k = 1, \dots, n$.

Now, we will consider a numerical sequence $\widehat{b} := (\widehat{b}_i)_i$ and a lower triangular matrix $\widehat{B}$, connected to b and B.

From the sequence $b := (b_i)_i$, we define the sequence $\widehat{b} := (\widehat{b}_i)_i$ as

$$\widehat{b}_0 = 0, \quad \sum_{k=1}^{n} b_{n,k} \widehat{b}_k = \delta_{n,1} \quad n = 1, 2, \dots. \tag{2.5}$$

The sequence $(\widehat{b}_k)_k$ is called the *conjugate sequence* of $(b_i)_i$.

Now, from the sequence $(\widehat{b}_k)_k$, we construct the matrices $\widehat{B}_N$ and $\widehat{B}_\infty$ as in (2.1)

Algorithm 2.1.2 (Construction of the conjugate matrix). For any $N \in \mathbb{N}$,

$$\begin{aligned}
&\text{for } l = 0, 1, \dots, N; \\
&\quad \widehat{b}_{l,0} = \delta_{l,0}; \\
&\quad \widehat{b}_{l,1} = \widehat{b}_l; \\
&\quad\quad \text{for } k = 2, \dots, l; \\
&\quad\quad \widehat{b}_{l,k} = \frac{1}{k} \sum_{i=1}^{l-k+1} \binom{l}{i} \widehat{b}_{i,1} \widehat{b}_{l-i,k-1}; \\
&\quad\quad \text{for } k = l+1, \dots, N; \\
&\quad\quad \widehat{b}_{l,k} = 0,
\end{aligned} \tag{2.6}$$

and we define the matrices $\widehat{B}_N := (\widehat{b}_{i,k})_{i,k=0,\dots,N}$ and $\widehat{B}_\infty \equiv \widehat{B} := (\widehat{b}_{l,k})_{l,k\geq 0}$. These are binomial-type matrices.

The matrices $\widehat{B}_N$ and $\widehat{B}$ are called *conjugate matrices*, respectively, of B_N and B.

Remark 2.4. If the numerical sequence $(\widehat{b}_i)_i$ with $\widehat{b}_0 = 0$ and $\widehat{b}_1 \neq 0$, is known, we can determine the conjugate sequence $(b_i)_i$ by

$$b_0 = 0, \quad \sum_{k=1}^{n} \widehat{b}_{n,k} b_k = \delta_{n,1} \quad n = 1, 2, \dots, \tag{2.7}$$

and the conjugate matrix $B = (b_{n,k})$ as in (2.1), and so the role of the sequences b and $\widehat{b}$ is interchangeable.

We consider the generating functions of $(\widehat{b}_i)_i$ and $(\widehat{b}_{i,k})_i$ as follows:
Putting

$$\overline{f}(x) = \sum_{i=0}^{\infty} \widehat{b}_i \frac{x^i}{i!},$$
$$\frac{1}{k!}(\overline{f}(x))^k = \sum_{i=0}^{\infty} \widehat{b}_{i,k} \frac{x^i}{i!},$$

we get

Proposition 2.2. *Let $f(x)$ and $\overline{f}(x)$ be the generating functions, respectively, of b and $\widehat{b}$. Then, the result is*

$$f(\overline{f}(x)) = \overline{f}(f(x)) = x. \tag{2.8}$$

Proof. From

$$\overline{f}(f(x)) = \sum_{i=1}^{\infty} \widehat{b}_i \frac{(f(x))^i}{i!}.$$

The proof follows applying (2.3) and (2.5).

Of course, if we consider the composition $f(\overline{f}(x))$, the proof follows from (2.6) and (2.7). □

We note that for (2.8), the series $\overline{f}(x)$ is the *compositional inverse* [167, pp. 120] of $f(x)$.

Remark 2.5. We note that (2.5) also follows from (2.8).

2.2 Appell-type matrix

Now, we consider another sequence of elements of $\mathbb{K}$, $a := (a_i)_i$ with $a_0 = 1$, and define a new lower triangular infinite matrix.

Algorithm 2.2.1 (Construction of infinite lower triangular matrix A). Let $N \in \mathbb{N}$:

$$\begin{aligned}
&\text{for } l = 0, 1, \ldots, N;\\
&\quad a_{l,0} = a_l;\\
&\qquad \text{for } k = 1, \ldots, l;\\
&\qquad\quad a_{l,k} = \binom{l}{k} a_{l-k};\\
&\qquad \text{for } k = l+1, \ldots, N;\\
&\qquad\quad a_{l,k} = 0;
\end{aligned} \tag{2.9}$$

Then, we set $A_N := (a_{l,k})_{l,k}$, $l = 0, \ldots, N$, $k = 0, \ldots, l$.

The matrix A_N is called an *Appell-type matrix of order N*.

For $N \to +\infty$, we get an infinite, lower triangular matrix of Appell type, and we denote it with A_∞ or, for simplicity, A. We note that A_N is the principal submatrix of A, of order N.

Just as in the binomial case, we can consider a sequence $(\widehat{a}_i)_i$ and a matrix $\widehat{A}$, connected, respectively, to a and A.

We define the numerical sequence $\widehat{a} := (\widehat{a}_i)_i$ by

$$\forall n \in \mathbb{N} \quad : \quad \sum_{k=0}^{n} \binom{n}{k} a_k \widehat{a}_{n-k} = \delta_{n,0}. \tag{2.10}$$

From the numerical sequence $(\widehat{a}_i)_i$, we can construct the matrix $\widehat{A}_N$ by Algorithm 2.2.1, that is,

Algorithm 2.2.2 (Construction of the conjugate matrix). For any $N \in \mathbb{N}$:

$$\begin{aligned}
&\text{for } l = 0, 1, \ldots, N;\\
&\quad \widehat{a}_{l,0} = \widehat{a}_l;\\
&\qquad \text{for } k = 1, \ldots, l;\\
&\qquad\quad \widehat{a}_{l,k} = \binom{l}{k} \widehat{a}_{l-k,0} = \binom{l}{k} \widehat{a}_{l-k};\\
&\qquad \text{for } k = l+1, \ldots, N;\\
&\qquad\quad \widehat{a}_{l,k} = 0;
\end{aligned} \tag{2.11}$$

We get $\widehat{A}_N := (\widehat{a}_{i,k})_{i,k=0,1,\ldots,N}$ and $\widehat{A}_\infty \equiv \widehat{A} := (\widehat{a}_{i,k})_{i,k\geq 0}$.

The sequence $\widehat{a}$ and the matrix $\widehat{A}$ are called, respectively, the *conjugate sequence* and the *conjugate matrix* of the sequence a and matrix A.

Remark 2.6. The matrix A in (2.9) can be decomposed as

$$A = D_1 T_a D_1^{-1}, \tag{2.12}$$

where

$$D_1 = \operatorname{diag}\{i!\}_{i=0,1,\dots}, \tag{2.13}$$

$$T_a = \begin{pmatrix} a_0 & & & & & \\ a_1 & a_0 & & & & \\ \frac{a_2}{2!} & a_1 & a_0 & & & \\ \frac{a_3}{3!} & \frac{a_2}{2!} & a_1 & a_0 & & \\ \frac{a_4}{4!} & \frac{a_3}{3!} & \frac{a_2}{2!} & a_1 & a_0 & \\ \vdots & & & & & \ddots \end{pmatrix}.$$

Given T_a, the Toeplitz matrix [148] of entries $t_{i,j} = \frac{a_{i-j}}{(i-j)!}$, from (2.12), we have that an Appell-type matrix is obtained by premultiplying and postmultiplying a Toeplitz matrix for a factorial diagonal matrix.

Proposition 2.3. *Let $g(x)$ be the generating function of the numerical sequence a, that is,*

$$g(x) = \sum_{i=0}^{\infty} a_i \frac{x^i}{i!}; \tag{2.14}$$

we have an invertible power series, resulting in

$$\frac{1}{g(x)} = \sum_{i=0}^{\infty} \hat{a}_i \frac{x^i}{i!}, \tag{2.15}$$

where $\hat{a}_i$ are defined in (2.10).

Proof. It is sufficient to verify that $g(x) \cdot \frac{1}{g(x)} = 1$. □

2.3 Sheffer-type matrix

For a Sheffer-type matrix, we consider two sequences of elements of $\mathbb{K}$: $(a_i)_i$ with $a_0 = 1$ and $(b_i)_i$ with $b_0 = 0$, $b_1 \neq 0$. Let $N \in \mathbb{N}$; we define an infinite lower triangular matrix as follows:

Algorithm 2.3.1 (Construction of infinite lower triangular matrix S). If $b_{i,j}$ are defined in (2.1) and $a_{i,j}$ in (2.9), $\forall N \in \mathbb{N}$, we set

$$\begin{aligned} &\text{for } i = 0, 1, \dots, N; \\ &\quad s_{i,0} = a_i; \\ &\quad\quad \text{for } j = 1, \dots, i; \\ &\quad\quad\quad s_{i,j} = \sum_{l=j}^{i} a_{i,l} b_{l,j}; \\ &\quad\quad \text{for } j = i+1, \dots, N; \\ &\quad\quad\quad s_{i,j} = 0. \end{aligned} \tag{2.16}$$

As a result, we get the matrix $S_N := (s_{i,j})_{i,j=0,\dots,N}$ and $S_\infty \equiv S := (s_{i,j})_{i,j\geq 0}$.

The matrices S_N and S are called *Sheffer-type matrices*, respectively, of order N and infinite. We note that S_N is the principal submatrix of S, of order N.

Remark 2.7. We note explicitly that, with the usual product of the matrix, the result is

$$S_N = A_N B_N \quad \forall N \in \mathbb{N},$$

that is, the Sheffer matrix S_N is the product of the Appell-type A_N and the binomial-type matrix B_N.

Moreover, for $b_1 = 1$ and $b_i = 0$, $i > 1$, the result is $S = A$, whereas for $a_0 = 1$ and $a_i = 0$, $i \geq 1$, we have $S = B$, that is, Appell-type or binomial-type matrices are Sheffer-type matrices, too.

Consequently, we observe—after easy calculations—that the ith column of S has as generating function the formal power series $\frac{1}{i!}g(x)f(x)^i$ (the first column being indexed by 0), that is,

$$\frac{1}{i!}g(x)f(x)^i = \sum_{k=0}^{\infty} s_{k,i}\frac{x^k}{k!}, \tag{2.17}$$

where $f(t)$ and $g(t)$ are defined in (2.2) and (2.14).

For the matrix S, we can define the conjugate. Let S be the matrix defined in (2.16), and let

$$\begin{cases} \widehat{s}_0 = a_0; \\ \widehat{s}_k = \sum_{i=1}^{k} a_i \widehat{b}_{k,i} \quad k = 1, 2, \dots, \end{cases} \tag{2.18}$$

where $\widehat{b}_{k,i}$ are defined in (2.5). From these coefficients, we can define the sequence $(g_i)_i$ by

$$\sum_{i=0}^{k} \binom{k}{i} \widehat{s}_i g_{k-i} = \delta_{k,0} \quad k = 0, 1, \dots. \tag{2.19}$$

Algorithm 2.3.2 (Construction of the conjugate matrix). With previous notations and definitions and for any $N \in \mathbb{N}$:

$$
\begin{aligned}
&\text{for } i = 0, 1, \ldots, N;\\
&\quad \widehat{s}_{i,0} = g_i;\\
&\quad \text{for } j = 1, \ldots, i;\\
&\qquad \widehat{s}_{i,j} = \sum_{l=0}^{i} \binom{i}{l} \widehat{s}_{l,0} \widehat{b}_{i-l,j} = \sum_{l=0}^{i} \binom{i}{l} g_l \widehat{b}_{i-l,j};\\
&\quad \text{for } j = i+1, \ldots, N;\\
&\qquad \widehat{s}_{i,j} = 0;
\end{aligned} \tag{2.20}
$$

Then we define $\widehat{S}_N := (\widehat{s}_{i,j})_{i,j=0,1,\ldots,N}$, $\widehat{S} := \widehat{S}_\infty = (\widehat{s}_{i,j})_{i,j\geq 0}$.

The matrices $\widehat{S}_N$ and $\widehat{S}$ are called, respectively, the *conjugate matrix* of S_N and S, $\widehat{S}_N$ is the principal submatrix of order N of $\widehat{S}$.

If we denote by $\widehat{A}^*$ the Appell-type matrix generated by the numerical sequence $(g_i)_i$ defined in (2.19), the matrix $\widehat{S}$ can be determined by Algorithm 2.3.1 from the matrices $\widehat{A}^*$ and $\widehat{B}$; therefore $\widehat{S}$ is also a Sheffer-type matrix.

Remark 2.8. We observe that:

1. the sequence $\widehat{s} = \{\widehat{s}_k\}_k$ is associated with the formal power series $g(\bar{f}(x))$, that is,

$$g(\bar{f}(x)) = \sum_{i=0}^{\infty} \widehat{s}_i \frac{x^i}{i!},$$

2. the sequence $g = \{g_i\}_i$ is associated with the formal power series $\frac{1}{g(\bar{f}(x))}$, that is,

$$\frac{1}{g(\bar{f}(x))} = \sum_{i=0}^{\infty} g_i \frac{x^i}{i!}.$$

Then it results that the kth column of $\widehat{S}$ is generated by the formal power series $\frac{1}{k!}(\frac{1}{g(\bar{f}(x))} \cdot (g(\bar{f}(x)))^k)$, that is,

$$\frac{1}{k!}\left(\frac{1}{g(\bar{f}(x))} \cdot (\bar{f}(x))^k\right) = \sum_{i=0}^{\infty} \widehat{s}_{i,k} \frac{x^i}{i!}. \tag{2.21}$$

As mentioned in [147], “The concept of representing columns of infinite matrices by formal power series is not new and goes back to Schur’s paper and Faber polynomials in 1945” [176].

Louis Shapiro et al. [177] introduced the concept of Riordan matrix; later this concept was generalized to the exponential Riordan matrix by many authors (see [19, 39, 103, 104, 137] and references therein).

In an exponential Riordan matrix, the jth column (being the first index with 0) has the generating function $\frac{1}{j!}g(x)(f(x))^j$ with $f(x)$ and $g(x)$ formal power series, respectively, δ-series and invertible. The exponential Riordan matrix generated by the formal power series $g(x), f(x)$ is denoted by $[g(x),f(x)]$.

Remark 2.9. Riordan matrices form a group called the Riordan group [39, 103], with the following properties:

- The multiplication is

$$[g(x),f(x)] * [h(x), l(x)] = [g(x)h(f(x)), l(f(x))]. \tag{2.22}$$

- The identity is

$$I = [1, x]. \tag{2.23}$$

- The inverse is

$$[g(x),f(x)]^{-1} = \left[\frac{1}{g(\bar{f}(x))}, \bar{f}(x)\right], \tag{2.24}$$

  where $\bar{f}(x)$ is the compositional inverse of $f(x)$.

If the considered numerical sequences $(a_i)_i$ and $(b_i)_i$ are finite, the corresponding generating functions are polynomials, and the product of matrices (2.22) becomes the usual matrix product. The infinite lower triangular matrices of Sheffer-type S and $\widehat{S}$ are *exponential Riordan matrices*, and precisely $S = [g(x),f(x)]$ and $\widehat{S} = [\frac{1}{g(\bar{f}(x))}, \bar{f}(x)]$.

Moreover, A and B are, also, exponential Riordan matrices, resulting in $A = [g(x), x]$ and $B = [1, f(x)]$. Likewise, $\widehat{B}$, $\widehat{A}$, A^*, $\widehat{A}^*$ are, respectively, the exponential Riordan matrices $[1, \bar{f}(x)]$, $[\frac{1}{g(x)}, x]$, $[g(\bar{f}(x)), x]$, and $[\frac{1}{g(\bar{f}(x))}, x]$.

Proposition 2.4. *The conjugate matrices $\widehat{B}$, $\widehat{A}$, $\widehat{A}^*$, $\widehat{S}$ are the inverse matrices respectively of B, A, A^*, S, that is,*

$$\widehat{B} = B^{-1}, \quad \widehat{A} = A^{-1}, \quad \widehat{S} = S^{-1}, \quad \widehat{A}^* = (A^*)^{-1}. \tag{2.25}$$

Moreover, the results are

$$\begin{aligned} &S = A \cdot B, \quad \widehat{S} = \widehat{A}^* \cdot \widehat{B} = \widehat{B} \cdot \widehat{A}, \\ &S_n = A_n \cdot B_n, \quad \widehat{S}_n = \widehat{A}_n^* \cdot \widehat{B}_n = \widehat{B}_n \cdot \widehat{A}_n \quad \forall n \in \mathbb{N}. \end{aligned} \tag{2.26}$$

Proof. The results follow from (2.22), (2.24) and Algorithms 2.1.2, 2.2.2, 2.3.2. □

Summarizing:

$S = [g(x), f(x)]$	$A = [g(x), x]$	$B = [1, f(x)]$	$A^* = [g(\overline{f}(x)), x]$
$S^{-1} = \widehat{S} = [\frac{1}{g(\overline{f}(x))}, \overline{f}(x)]$	$A^{-1} = \widehat{A} = [\frac{1}{g(x)}, x]$	$B^{-1} = \widehat{B} = [1, \overline{f}(x)]$	$(A^*)^{-1} = \widehat{A}^* = [\frac{1}{g(\overline{f}(x))}, x]$

2.4 Lidstone-type matrix

Let $(\alpha_{2i})_i$ be a sequence of elements of $\mathbb{K}$ with $\alpha_0 \neq 0$. We consider for any $N \in \mathbb{N}$ the lower triangular matrix, defined by the following algorithm:

Algorithm 2.4.1 (Odd Lidstone-type matrix). If we set for any $N \in \mathbb{N}$,

$$\begin{aligned}
&\text{for } i = 0, 1, \ldots, N;\\
&\quad o_{i,0} = \alpha_{2i};\\
&\quad\text{for } j = 1, \ldots, i;\\
&\qquad o_{i,j} = \binom{2i+1}{2j+1} \frac{\alpha_{2(i-j)}}{2(i-j)+1};\\
&\quad\text{for } j = i+1, \ldots, N;\\
&\qquad o_{i,j} = 0,
\end{aligned} \tag{2.27}$$

we define the matrix $O_N := (o_{i,j})_{i,j=0,1,\ldots,N}$ and for $N \to \infty$, $O_\infty := O = (o_{i,j})_{i,j\geq 0}$.

The matrices O_N and O are called *odd Lidstone-type matrices*, respectively, of order N and infinite. The matrix O_N is the principal submatrix of O, of order N.

We consider, now, the numerical sequence $(\beta_{2i})_i$, defined by

$$\sum_{j=0}^{i} \frac{\beta_{2j}\alpha_{2(i-j)}}{(2j+1)!(2(i-j)+1)!} = \delta_{i,0}, \quad i = 0, 1, \ldots. \tag{2.28}$$

Then, from the sequence $(\beta_{2i})_i$, we can construct the matrix $\widehat{O}$ as in Algorithm 2.4.1. Hence, the matrix $\widehat{O}$ has entries defined by

$$\begin{cases}
\widehat{o}_{i,0} = \beta_{2i}, & i = 0, 1, \ldots,\\
\widehat{o}_{i,j} = \binom{2i+1}{2j+1}\frac{\beta_{2(i-j)}}{2(i-j)+1}, & i = 1, 2, \ldots, \quad j = 1, \ldots, i,\\
\widehat{o}_{i,j} = 0, & i < j.
\end{cases} \tag{2.29}$$

The matrices O and $\widehat{O}$ are called *conjugate odd-Lidstone-type matrices*. $\widehat{O}_N$ is the principal submatrix of $\widehat{O}$, of order N.

Proposition 2.5. *With previous notations and hypothesis, if we set*

$$l(t) = \sum_{k=0}^{\infty} \alpha_{2k} \frac{t^{2k}}{(2k+1)!} \tag{2.30}$$

and

$$\frac{1}{l(t)} = \sum_{k=0}^{\infty} \beta_{2k} \frac{t^{2k}}{(2k+1)!}, \tag{2.31}$$

the result is

$$l(t) \cdot \frac{1}{l(t)} = 1.$$

Proof. The proof follows by easy calculations and from (2.28). □

The power series $l(t)$ is called the *generating function* of the numerical sequence $(\alpha_{2k})_k$, and, in analogy, $\frac{1}{l(t)}$ is the *generating function* of the sequence $(\beta_{2k})_k$.

Proposition 2.6. *For an odd Lidstone-type matrix O, we get the decomposition*

$$O = D_1 T_{2\alpha} D_1^{-1}, \tag{2.32}$$

where

$$D_1 \text{ is the diagonal matrix } D_1 = \operatorname{diag}\{(2i+1)!\ i = 0, 1, \ldots\},$$
$$T_{2\alpha} \text{ is the Toeplitz matrix with entries } t_{i,j} = \frac{\alpha_{2(i-j)}}{(2(i-j)+1)!}.$$

Moreover, O is invertible, and its result is

$$O^{-1} = \widehat{O}.$$

Proof. From (2.32), the inverse matrix has the decomposition

$$O^{-1} = D_1 T_{2\beta} D_1^{-1}, \tag{2.33}$$

where $T_{2\beta}$ is the Toeplitz matrix [201] with entries

$$\widehat{t}_{i,j} = \frac{\beta_{2(i-j)}}{(2(i-j)+1)!}.$$

Hence, $O^{-1} = \widehat{O}$. □

Now, we want to define another Lidstone-type matrix.

Given a sequence of elements of $\mathbb{K}$, $(\gamma_{2i})_i$ with $\gamma_0 \neq 0$, for any $N \in \mathbb{N}$, we consider the lower triangular matrix defined by the following algorithm:

Algorithm 2.4.2 (Even Lidstone-type matrix). For any $N \in \mathbb{N}$:

$$
\begin{aligned}
&\text{for } i = 0, 1, \ldots, N; \\
&\quad e_{i,0} = \gamma_{2i}; \\
&\quad \text{for } j = 1, \ldots, i; \\
&\qquad e_{i,j} = \binom{2i}{2j} \gamma_{2(i-j)}; \\
&\quad \text{for } j = i+1, \ldots, N; \\
&\qquad e_{i,j} = 0;
\end{aligned}
\tag{2.34}
$$

We define the matrices $E_N := (e_{i,j})_{i,j=0,1,\ldots,N}$ and for $N \to \infty$, $E_\infty := E = (e_{i,j})_{i,j\geq 0}$.

The matrices E_N and E are called, respectively, *even Lidstone-type matrices* of order N and infinite.

The matrix E_N is the principal submatrix of E, of order N.

We consider the numerical sequence $(\delta_{2i})_i$, defined by

$$
\sum_{j=0}^{i} \frac{\gamma_{2j}\delta_{2(i-j)}}{(2j)!(2(i-j))!} = \delta_{i,0}, \quad i = 0, 1, \ldots . \tag{2.35}
$$

Then, from the sequence (δ_{2i}), we can define the matrix $\widehat{E}$ by the Algorithm 2.4.2, that is,

$$
\begin{cases}
\widehat{e}_{i,0} = \delta_{2i} & i = 0, 1, \ldots \\
\widehat{e}_{i,j} = \binom{2i}{2j}\delta_{2(i-j)} & i = 1, 2, \ldots, \quad j = 1, \ldots, i \\
\widehat{e}_{i,j} = 0 & i < j.
\end{cases}
\tag{2.36}
$$

The matrices E and $\widehat{E}$ are called *conjugate even Lidstone-type matrices*.

Proposition 2.7. *If we set*

$$
h(t) = \sum_{i=0}^{\infty} \gamma_{2i} \frac{t^{2i}}{(2i)!}, \tag{2.37}
$$

and

$$
\frac{1}{h(t)} = \sum_{i=0}^{\infty} \delta_{2i} \frac{t^{2i}}{(2i)!}, \tag{2.38}
$$

the result is

$$
h(t) \cdot \frac{1}{h(t)} = 1.
$$

Proof. It follows from (2.35). □

The power series $h(t)$ is called the *generating function* of the numerical sequence $(y_{2k})_k$ and, in analogy, $\frac{1}{h(t)}$ is the *generating function* of the sequence $(\delta_{2k})_k$.

Proposition 2.8. *For an even Lidstone-type matrix E, we get the decomposition*

$$E = D_2 \overline{T}_{2y} D_2^{-1}, \tag{2.39}$$

where

$$D_2 \text{ is the diagonal matrix } : D_2 = \operatorname{diag}\{(2i)!\, i = 0, 1, \ldots\},$$
$$\overline{T}_{2y} \text{ is the Toeplitz matrix with entries } \overline{t}_{i,j} = \frac{y_{2(i-j)}}{(2(i-j))!}.$$

Moreover, the matrix E is invertible and results in

$$E^{-1} = \widehat{E}.$$

Proof. The proof follows after easy calculations, according to previous notations and known results. □

Finally, there is an interesting relationship between Appell- and Lidstone-type matrices.

Remark 2.10. If (α_i) is a numerical sequence with $\alpha_0 \neq 0$, from Algorithm 2.2.1, we can construct an Appell-type matrix $A_{(\alpha)}$. If we consider the subsequence $(\alpha_{2i})_i$, we can get both odd and even Lidstone-type matrices; in fact, we get $O_{(2\alpha)}$ from Algorithm 2.4.1 and $E_{(2\alpha)}$ from Algorithm 2.4.2.

$A_{(\alpha)}$, $O_{(2\alpha)}$, and $E_{(2\alpha)}$ are called *Appell- and Lidstone-type associated matrices.*

Vice versa from a sequence $(\alpha_{2i})_i$, $\alpha_0 \neq 0$, we can consider infinite sequences $(\overline{\alpha}_i)_i$ with $\overline{\alpha}_{2i} = \alpha_{2i}$. Then with an odd and even Lidstone-type matrix, we can associate infinite Appell-type matrices.

In Figure 2.1, we show Remark 2.10.

2.5 Production matrix

Now, we introduce the new concept of *production matrix*.

Definition 2.1 (Production matrix, [77]). For an invertible infinite lower triangular matrix A, its production matrix (also called Stieltjets matrix) is the matrix R, such as

$$R = A^{-1}\overline{A}, \tag{2.40}$$

where $\overline{A}$ is the matrix A, with its first row removed.

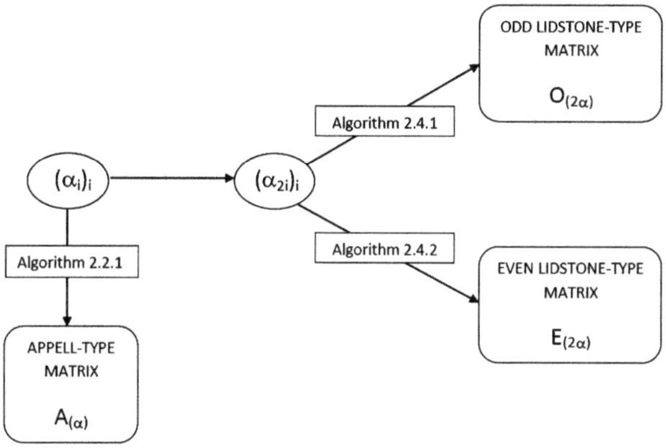

Figure 2.1: Appell- and Lidstone-type associated matrices.

Equivalently, if $D = (\delta_{i+1,j})_{i,j\geq 0}$, given $\delta_{i,j}$, the Kronecker symbol, the production matrix is defined by

$$AR = DA. \tag{2.41}$$

The production matrix is a Hessemberg matrix.

For an explicit calculation of the entries of the matrix R, we have

Proposition 2.9 (Construction of production matrix). *Let $A = (a_{i,j})$ an invertible infinite matrix, and let $A^{-1} = (a^{-1}_{i,j})$ be its inverse; if*

$$r_{i,j} = \sum_{k=0}^{i} a_{i,k} a^{-1}_{k+1,j} \quad i,j = 0,1,\ldots, \tag{2.42}$$

then the matrix $R = (r_{i,j})$ is the production matrix of A^{-1}.

2.6 Summary

In this introductive part, we have defined, by means of two numerical sequences, particular classes of infinite, lower, and triangular matrices, called binomial-type, Appell-type, Sheffer-type, and Lidstone-type matrices.

To the numerical sequences, we have associated formal power series, called generating functions; one of these is invertible, that is,

$$g(x) = \sum_{i=0}^{\infty} a_i \frac{x^i}{i!} \quad \text{with} \quad a_0 \neq 0,$$

and the other is a δ-series [167, p. 4], that is,

$$f(x) = \sum_{i=0}^{\infty} b_i \frac{x^i}{i!} \quad \text{with} \quad b_0 = 0, \quad b_1 \neq 0.$$

The classes of binomial- and Appell-type matrices are included in the Sheffer type.

Sheffer matrices are exponential Riordan matrices, that is, the jth column (being the first indexed with 0) has as its generating function the formal power series $\frac{1}{j!}g(x)(f(x))^j$. The set of exponential Riordan matrices is a group, so a Sheffer matrix is invertible. Lidstone-type matrices have been introduced in this context for their close connection with Appell-type matrices.

We have also defined the concept of conjugate matrix, and we have proved that it coincides with the inverse matrix for these classes. Finally, the concept of production matrix has been introduced.

Part II: **Polynomial sequences of binomial type**

It remains a mystery why so many polynomial sequences occurring in various mathematical circumstances turn out to be binomial type.
S. Roman, G. C. Rota (1978)

3 Binomial polynomial sequences

Abstract: The purpose of this chapter is to present a theory of polynomial sequences of binomial-type, based on matrix calculus, recurrence, and determinantal forms. We consider also some connections with previous approaches, such as generating functions and operators methods. An interesting differential relation is presented, which will establish a close link between binomial, Appell, and Sheffer sequences.

3.1 Introduction

"The notion of Binomial-type polynomial sequences [167] (b. p. s. in the following) goes back to E. T. Bell [21, 22]" and A. C. Aitken [11, 128]. "Steffens was the first to observe that the sequences associated to delta operators, in the same way as D is associated to x^n, are the binomial-type; but he did not notice the converse of this fact, which was first stated and proved by Mullin and Rota" [167].

Polynomial sequences of binomial-type turn up in a large variety of mathematical problems: combinatorial analysis and probability [32, 78, 80], interpolation and linear operators approximation [10, 135]. In 1970, Mullin and Rota [142] gave the first systematic theory, using operators methods instead of the less efficient generating function methods [179] that had been exclusively used until then. Garsia [93] observed: "Unfortunately, the notations and the proofs in that very original paper in some instances leave something to be desired, and even tend to obscure the remarkable simplicity and beauty of the results." An algebraic approach to Rota–Mullin theory has been considered in [91].

In the following, we present an approach based on matrix calculus, recurrence, and determinantal formulas. Necessary links with previous theories are given.

3.2 Definition and characterizations

Let P be a binomial-type matrix (2.1), that is, there exists a sequence $b := (b_n)_{n\in\mathbb{N}}$ of elements of $\mathbb{K}$ such that $b_0 = 0$, $b_1 \neq 0$, and if we set

$$\begin{cases} p_{n,0} = \delta_{n,0}, & n = 0, 1, \dots, \\ p_{n,1} = b_n, & n = 1, 2, \dots, \\ p_{n,k} = \frac{1}{k}\sum_{i=1}^{n-k+1}\binom{n}{i}p_{i,1}p_{n-i,k-1}, & n = 2, 3, \dots, \quad k = 2, \dots, n, \\ p_{n,k} = 0, & k > n, \end{cases} \tag{3.1}$$

we have the matrix $P := (p_{i,k})_{i,k=0,1,\dots}$.

https://doi.org/10.1515/9783110652925-003

Definition 3.1. The polynomial sequence $\{p_n\}_n$ defined by

$$\begin{cases} p_0(x) = 1, \\ p_1(x) = p_{1,0} + p_{1,1}x, \\ \cdots\cdots\cdots \\ p_k(x) = p_{k,0} + p_{k,1}x + \cdots + p_{k,k}x^k, \\ \cdots\cdots\cdots \end{cases} \tag{3.2}$$

is called a *binomial polynomial sequence* (b. p. s. in the following) with matrix P.

Remark 3.1. We note that the elements $p_{l,k}$ can be considered as Bell polynomials [166, pp. 82], [22], but their connection with b. p. s. [139] will not be debated in the following.

The following characterization explains the construction of binomial-type matrix P, as in Algorithm 2.1.1.

Theorem 3.1 (Differential relation). *Let $\{p_n\}_n$ be a polynomial sequence. It is a b. p. s. if and only if there exists a numerical sequence $b := (b_i)_i$ with $b_0 = 0$ and $b_1 \neq 0$ such that*

$$\forall n = 0, 1, \dots, \qquad \begin{cases} p_n'(x) = \sum_{i=1}^{n} \binom{n}{i} b_i p_{n-i}(x) = \sum_{i=0}^{n-1} \binom{n}{i} b_{n-i} p_i(x), \\ p_n(0) = 0, \quad n \geq 1, \quad p_0(x) = 1. \end{cases} \tag{3.3}$$

Proof. Let $\{p_n\}_n$ be a b. p. s. Hence, there exists a numerical sequence $(b_i)_i$ with $b_0 = 0$ and $b_1 \neq 0$ such that $p_n(x) = \sum_{k=0}^{n} p_{n,k}x^k$, where $p_{n,k}$ are defined in (3.1). Then we have

$$p_n'(x) = \sum_{k=1}^{n} p_{n,k} k x^{k-1} = \sum_{k=1}^{n} \left(\sum_{i=1}^{n-k+1} \binom{n}{i} b_i p_{n-i,k-1} \right) x^{k-1}. \tag{3.4}$$

From (3.4), it follows:

$$p_n'(x) = \sum_{i=1}^{n} \binom{n}{i} b_i p_{n-i}(x).$$

With the reverse procedure, after integration, the opposite implication follows. □

Remark 3.2. Relation (3.4) contains the recurrence form of Algorithm 2.1.1, that is, the recurrence relation (3.1) is equivalent to the differential relation (3.3).

Proposition 3.1. *A b. p. s. $\{p_n\}_n$ is nonnegative in $[0, +\infty)$ if and only if the numerical sequence (b_i) is nonnegative.*

Proof. If (b_i) is nonnegative, that is, $b_i \geq 0$, $\forall i \in \mathbb{N}$, from (3.1) and (3.2) the result follows. Vice versa, if $p_n(x) \geq 0$, $\forall x \in [0, +\infty)$, since $p_n(0) = 0$, it results in $p_n'(0) \geq 0$. But from (3.3), $p_n'(0) = b_n$, and this completes the proof. □

Let $\widehat{P} = (\widehat{p}_{i,k})$ be the conjugate matrix of P as in Algorithm 2.1.2. $\widehat{P}$ being a binomial-type matrix, we can consider the b.p.s with matrix $\widehat{P}$:

$$\begin{cases} \widehat{p}_0(x) = 1, \\ \widehat{p}_1(x) = \widehat{p}_{1,0} + \widehat{p}_{1,1}x, \\ \dots\dots\dots \\ \widehat{p}_k(x) = \widehat{p}_{k,0} + \widehat{p}_{k,1}x + \dots + \widehat{p}_{k,k}x^k, \\ \dots\dots\dots \end{cases} \tag{3.5}$$

The sequence $\{\widehat{p}_k\}_k$ is called the *conjugate binomial polynomial sequence* of $\{p_n\}_n$.

If $\{p_n\}_n$ and $\{q_n\}_n$ are two polynomial sequences with $p_n(x) = \sum_{k=0}^n p_{n,k}x^k$ and $q_n(x) = \sum_{k=0}^n q_{n,k}x^k$, their umbral composition [166, p. 41] is

$$p_n \circ q_n = p_n(q_n(x)) = \sum_{k=0}^{n} p_{n,k}q_k(x); \quad \sum_{k=0}^{n} q_{n,k}p_k(x) = q_n(p_n(x)) = q_n \circ p_n. \tag{3.6}$$

Then we get

Proposition 3.2. *If $\{p_n\}_n$ and $\{\widehat{p}_n\}_n$ are conjugate b. p. s., we have*

$$p_n(\widehat{p}_n(x)) = \widehat{p}_n(p_n(x)) = x^n, \quad n = 1, \dots. \tag{3.7}$$

Proof. The result follows from properties of the matrices P and $\widehat{P}$ associated with $\{p_n\}_n$ and $\{\widehat{p}_n\}_n$, respectively. □

Corollary 3.1. *Let $\mathcal{B} = \{\{p_n\}_n \mid \{p_n\}_n$ is a b.p.s.$\}$. The structure $(\mathcal{B}, \circ)$, where $\circ$ is the umbral composition, is the group in which*
- *the identity is $\{x^n\}_n$;*
- *the inverse of $\{p_n\}_n$ is its conjugate $\{\widehat{p}_n\}_n$.*

We observe that, by Corollary 3.1, the polynomial sequence $\{x^n\}_n$ is the b. p. s. with numerical sequence $b_0 = 0$, $b_1 = 1$, $b_i = 0$, $i \geq 2$. For this p. s., the binomial identity is known:

$$(x + y)^n = \sum_{k=0}^{n} \binom{n}{k} x^k y^{n-k}. \tag{3.8}$$

Now we will extend this identity to a generic b. p. s.

Theorem 3.2 (Binomial identity). *A polynomial sequence $\{p_n\}_n$ is a b. p. s. if and only if the following relation holds:*

$$p_n(x + y) = \sum_{k=0}^{n} \binom{n}{k} p_k(x)p_{n-k}(y). \tag{3.9}$$

Proof. Let $p_n(x) = \sum_{k=0}^{n} p_{n,k}x^k$. If (3.9) holds, we have the following relation for the coefficients [167, Prop. 4.3]:

$$\binom{i+j}{i} p_{n,i+j} = \sum_{l=0}^{n} \binom{n}{l} p_{l,1}p_{n-l,j}, \quad i,j = 0,\dots,n, \tag{3.10}$$

which for $i = 1$ and $j = k-1$ leads to

$$p_{n,k} = \frac{1}{k} \sum_{l=0}^{n-k+1} \binom{n}{l} p_{l,1}p_{n-l,k-1}. \tag{3.11}$$

Then the result follows by Definition 3.1.

It is vice versa that if $\{p_n\}_n$ is a b. p. s., (3.11) holds, and then the (3.9) follows, after calculations, following the identity principle for polynomials. □

Corollary 3.2. *For a b. p. s. $\{p_n\}_n$, we have the discrete orthogonality*

$$\sum_{k=0}^{n} \binom{n}{k} p_n(x)p_{n-k}(-x) = 0, \quad \forall x \in \mathbb{R}. \tag{3.12}$$

Proof. The result follows from Theorem 3.2 for $y = -x$ and by (3.3). □

Corollary 3.3 ([167, p. 102]). *From the binomial identity, by iterations, the multimonomial identity holds:*

$$p_n(x_1 + x_2 + \cdots + x_n) = \sum \binom{n}{j_1,\dots,j_k} p_{j_1}(x)\cdots p_{j_k}(x), \tag{3.13}$$

where the sum is over all k-tuples of nonnegative integers $(j_1,\dots,j_k)$ for which $j_1 + \cdots + j_k = n$.

Remark 3.3. After Theorem 3.2, Definition 3.1 and the definition of b. p. s. given by Roman and Rota in [167, p. 102] are equivalent.

It is possible to characterize a b. p. s. with a generating function [163, pp. 129]. In fact, let $\{p_n\}_n$ be a b. p. s. By Definition 3.1, there exists a numerical sequence $(b_n)_n$ with $b_0 = 0$ and $b_1 \neq 0$, which gives the formal power series

$$f(t) = \sum_{n=0}^{\infty} b_n \frac{t^n}{n!}. \tag{3.14}$$

Then we get:

Theorem 3.3 (Generating function). *A polynomial sequence $\{p_n\}_n$ is a b. p. s. if and only if there exists a formal power series $f(t)$ as in (3.14), such that*

$$\exp(xf(t)) = \sum_{n=0}^{\infty} p_n(x)\frac{t^n}{n!}. \tag{3.15}$$

Proof. If $\{p_n\}_n$ is a b. p. s., relation (3.15) follows by Definition 3.1 and Proposition 2.1, after standard calculations.

Vice versa, from (3.15) and (3.14), we get (3.1)—and hence—we have a b. p. s. by Definition 3.1. □

The function

$$G(x,t) = \exp(xf(t)) \tag{3.16}$$

is called the *generating function* of b. p. s. $\{p_n\}_n$.

Corollary 3.4. *If $f(t)$ defined in (3.14) gives the b. p. s. $\{p_n\}_n$, then $f(-t)$ generates the sequence $\{(-1)^n p_n\}_n$.*

Proof. From the generating function, we have

$$\exp(x(f(-t))) = \sum_{n=0}^{\infty} p_n(x)\frac{(-t)^n}{n!} = \sum_{n=0}^{\infty} (-1)^n p_n(x)\frac{t^n}{n!},$$

and from this the result follows. □

Proposition 3.3. *With the previous notations, the following identity holds:*

$$\left(\exp(xf(t)) = \sum_{n=0}^{\infty} p_n(x)\frac{t^n}{n!}\right) \Leftrightarrow \left(p_n(x+y) = \sum_{k=0}^{n}\binom{n}{k} p_k(x)p_{n-k}(y)\right). \tag{3.17}$$

Proof. The result follows from known properties of exponential functions and the Cauchy product of series. But it can also follow from combining Theorem 3.3 and 3.2. □

Remark 3.4. The differential relation (3.3) can also be proved by a generating function. In fact, after differentiation with respect to the variable x and applying the Cauchy product of series, we get

$$p_n'(x) = \sum_{l=1}^{n}\binom{n}{l} p_{l,1}p_{n-l}(x),$$

that is, (3.3).

Moreover, by repeated differentiations with respect to the variable x, we get

$$p_n^{(k)}(x) = \sum_{l=k}^{n}\binom{n}{l} k!p_{l,k}p_{n-l}(x) \quad k = 1,\dots,n. \tag{3.18}$$

3.3 Recurrence relations

As disclosed in the Preface, we will determine recurrence relations for polynomial sequences of binomial type.

Theorem 3.4 (First recurrence relation). *Let $\{p_n\}$ and $\{\widehat{p}_n\}$ be conjugate b. p. s. Then we have*

$$p_n(x) = \frac{1}{\widehat{p}_{n,n}}\left(x^n - \sum_{k=0}^{n-1} \widehat{p}_{n,k} p_k(x)\right), \tag{3.19}$$

$$\widehat{p}_n(x) = \frac{1}{p_{n,n}}\left(x^n - \sum_{k=0}^{n-1} p_{n,k} \widehat{p}_k(x)\right), \tag{3.20}$$

where $p_{n,k}$ and $\widehat{p}_{n,k}$ are the coefficients of $p_n(x)$ and $\widehat{p}_n(x)$ on the canonical basis.

Proof. For conjugate b. p. s., we have (3.7), which means

$$\sum_{k=0}^{n} \widehat{p}_{n,k} p_k(x) = x^n = \sum_{k=0}^{n} p_{n,k} \widehat{p}_k(x).$$

These relations are equivalent to (3.19) and (3.20). □

Corollary 3.5. *The sets $\{p_0, p_1, \dots, p_n\}$ and $\{\widehat{p}_0, \widehat{p}_1, \dots, \widehat{p}_n\}$ are the basis of $\mathcal{P}_n$, for any $n \in \mathbb{N}$.*

Let $\{p_n\}_n$ and $\{\widehat{p}_n\}_n$ be conjugate b. p. s.; we consider the vectors:

$$\begin{aligned} \overline{P}(x) &= [p_0(x), p_1(x), \dots, p_n(x), \dots]^T, \\ \overline{\widehat{P}}(x) &= [\widehat{p}_0(x), \widehat{p}_1(x), \dots, \widehat{p}_n(x), \dots]^T, \\ X(x) &= [1, x, \dots, x^n, \dots]^T, \end{aligned} \tag{3.21}$$

and, for any $n \in \mathbb{N}$,

$$\begin{aligned} \overline{P}_n(x) &= [p_0(x), p_1(x), \dots, p_n(x)]^T, \\ \overline{\widehat{P}}_n(x) &= [\widehat{p}_0(x), \widehat{p}_1(x), \dots, \widehat{p}_n(x)]^T, \\ X_n(x) &= [1, x, \dots, x^n]^T. \end{aligned}$$

Then we have:

Proposition 3.4. *The following relations hold:*

$$\overline{P}(x) = PX(x), \quad X(x) = \widehat{P}\overline{P}(x), \tag{3.22}$$

$$\overline{\widehat{P}}(x) = \widehat{P}X(x), \quad X(x) = P\overline{\widehat{P}}(x), \tag{3.23}$$

$$\overline{P}(x) = P^2\overline{\widehat{P}}(x), \quad \overline{\widehat{P}}(x) = \widehat{P}^2\overline{P}(x), \tag{3.24}$$

and, for any $n \in \mathbb{N}$,

$$\begin{aligned} \overline{P}_n(x) &= P_n X_n(x), & X_n(x) &= \widehat{P}_n \overline{P}_n(x), \\ \overline{\widehat{P}}_n(x) &= \widehat{P}_n X_n(x), & X_n(x) &= P_n \overline{\widehat{P}}_n(x), \\ \overline{P}_n(x) &= P_n^2 \overline{\widehat{P}}_n(x), & \overline{\widehat{P}}_n(x) &= \widehat{P}_n^2 \overline{P}_n(x). \end{aligned}$$

Proof. The results follow from the above definitions, after easy calculations. □

Remark 3.5. The proof of Theorem 3.4 also follows from Proposition 3.4.

Proposition 3.5 (Derivation matrix). *$\forall n \in \mathbb{N}$ let:*

$$P'_n(x) = [p'_0(x), p'_1(x), \dots, p'_n(x)]^T,$$

$$\mathcal{D}P_n := (p^*_{l,k}) \quad \text{with} \quad p^*_{l,k} = \binom{l}{k} b_{l-k}, k = 0, \dots, l,\ l = 0, 1, \dots\ n.$$

Then, the results is

$$P'_n(x) = \mathcal{D}P_n \cdot P_n(x).$$

Proof. The result follows from (3.3) and above notations. □

The matrix $\mathcal{D}P$ will be called the *derivation matrix* for b. p. s.

Now we get:

Theorem 3.5 (Second recurrence relation). *Let $\{p_n\}_n$ be a polynomial sequence. It is a b. p. s. if and only if there exists a numerical sequence (b_i) with $b_0 = 0$ and $b_1 \neq 0$ such that the following relation holds:*

$$\widehat{c}_0 p_{n+1}(x) = -(n\widehat{c}_1 - x)p_n(x) - \sum_{k=2}^{n} \binom{n}{k} \widehat{c}_k p_{n-k+1}(x), \quad n = 1, 2, \dots, \tag{3.25}$$

where

$$\forall\, n \in \mathbb{N}, \quad \sum_{k=0}^{n} \binom{n}{k} \widehat{c}_k b_{n-k+1} = \delta_{n,0} \tag{3.26}$$

with initial conditions $\widehat{c}_0 = 1$, $b_1 = 1$, $p_0(x) = 1$, $p_1(x) = x$.

Proof. Let $\{p_n\}_n$ be the b. p. s. with the numerical sequence $(b_i)_i$, where $b_0 = 0$ and $b_1 \neq 0$, then we have the generating function (3.15), that is,

$$\exp(xf(t)) = \sum_{n=0}^{\infty} p_n(x)\frac{t^n}{n!}$$

with $f(t) = \sum_{i=0}^{\infty} b_i \frac{t^i}{i!}$.

The partial derivative of the generating function respect to the variable t gives

$$f'(t)x \exp(xf(t)) = \sum_{n=0}^{\infty} p_{n+1}(x)\frac{t^n}{n!}. \tag{3.27}$$

If $c_i = b_{i+1}$, $i = 0, 1, \dots$, we have $c_0 \neq 0$. Then the power series $f'(t) = \sum_{i=0}^{\infty} c_i \frac{t^i}{i!}$ is invertible, and it results in

$$\frac{1}{f'(t)} = \sum_{i=0}^{\infty} \widehat{c}_i \frac{t^i}{i!}$$

with $\widehat{c}_n$, $n = 0, 1, \dots,$ defined in (2.18) as

$$\sum_{i=0}^{n} \binom{n}{i} c_{n-i} \widehat{c}_i = \delta_{n,0}. \tag{3.28}$$

Writing (3.27) in the equivalent form

$$x \sum_{n=0}^{\infty} p_n(x) \frac{t^n}{n!} = \sum_{n=0}^{\infty} p_{n+1}(x) \frac{t^n}{n!} \cdot \sum_{n=0}^{\infty} \widehat{c}_n \frac{t^n}{n!},$$

with the use of the Cauchy power series product, we get

$$x \sum_{n=0}^{\infty} p_n(x) \frac{t^n}{n!} = \sum_{n=0}^{\infty} \left(\sum_{k=0}^{n} \binom{n}{k} p_{k+1}(x) \widehat{c}_{n-k} \right) \frac{t^n}{n!}.$$

From this we have

$$x p_n(x) = \sum_{k=0}^{n} \binom{n}{k} p_{k+1}(x) \widehat{c}_{n-k},$$

which completes the proof.

Vice versa, from (3.25), taking advantage of the identity principle for polynomials, we get (3.1) and, hence, a b. p. s. □

Remark 3.6. We note that the linear system (3.28) has the solution $\forall n = 1, 2, \dots,$

$$\widehat{c}_n = \frac{(-1)^i}{b_1^{n+1}} \times \begin{vmatrix} c_1 & c_0 & 0 & \cdots & 0 & 0 & \cdots & 0 \\ c_2 & \binom{2}{1} c_1 & c_0 & \cdots & 0 & 0 & \cdots & 0 \\ \vdots & & & \ddots & & & & \vdots \\ \vdots & & & & & & & \vdots \\ c_{n-1} & \binom{n-1}{1} c_{n-2} & \binom{n-1}{2} c_{n-3} & \cdots & \binom{n-1}{i-1} c_{n-i} & \binom{n-1}{i+1} c_{n-i-2} & \cdots & c_0 \\ c_n & \binom{n}{1} c_{n-1} & \binom{n}{2} c_{n-2} & \cdots & \binom{n}{i-1} c_{n-i+1} & \binom{n}{i+1} c_{n-i-1} & \cdots & \binom{n}{n-1} c_{n-1} \end{vmatrix}.$$

In particular,

$$\widehat{c}_0 = \frac{1}{b_1}, \quad \widehat{c}_1 = -\frac{b_2}{b_1^2}, \quad \widehat{c}_2 = \frac{1}{b_1^3}(2b_2^2 - b_3 b_1).$$

Remark 3.7. If the recurrence relation (3.25) is of three terms, that is,

$$\sum_{k=0}^{n-3} \binom{n}{k} \widehat{c}_{n-k} p_{k+1}(x) = 0 \quad \forall x,$$

with some suitable hypothesis on the sequence $(b_i)_i$, the b. p. s. is orthogonal [90, 191].

Hence, an orthogonal b. p. s. has the recurrence relation

$$p_{n+1}(x) = \left(n\frac{b_2}{b_1} + xb_1\right)p_n(x) - \frac{(2b_2^2 - b_3b_1)}{2b_1^3}n(n-1)p_{n-1}(x) \tag{3.29}$$

with $\frac{2b_2^2 - b_3b_1}{2b_1^3} > 0$.

Theorem 3.6 (Third recurrence relation). *Let $\{p_n\}_n$ be a polynomial sequence. It is a b. p. s. if and only if there exists a numerical sequence $(b_i)_i$ with $b_0 = 0$ and $b_1 \neq 0$ such that the following relation holds:*

$$p_{n+1}(x) = x\sum_{k=0}^{n}\binom{n}{k} b_{n-k+1}p_k(x). \tag{3.30}$$

Proof. The result follows from (3.27), applying the Cauchy product to the left-hand side. □

Remark 3.8. The recurrence relation (3.25) is proved in [215, Corollary 4.19] only as a necessary condition, by a more complex procedure.

3.4 Determinantal forms

Let $\{p_n\}_{n\in\mathbb{N}}$ be a b. p. s.. We will give a new characterization in a determinantal form.

Theorem 3.7 (First determinantal form, [65, 66]). *Two polynomial sequences $\{p_n\}_{n\in\mathbb{N}}$ and $\{\widehat{p}_n\}$ are conjugate b. p. s. if and only if there exists a numerical sequence $(b_n)_{n\in\mathbb{N}}$ with $b_0 = 0$ and $b_1 \neq 0$ such that if $\{\widehat{b}_n\}_{n\in\mathbb{N}}$ and $\widehat{p}_{n,k}$ are defined in (2.5) and (2.6), the following relations hold:*

$$\begin{cases} p_0(x) = 1, \\ p_n(x) = \dfrac{(-1)^{n+1}}{\prod_{k=0}^{n}\widehat{p}_{k,k}} \begin{vmatrix} x & x^2 & \cdots & x^{n-1} & x^n \\ \widehat{p}_{1,1} & \widehat{p}_{2,1} & \cdots & \widehat{p}_{n-1,1} & \widehat{p}_{n,1} \\ 0 & \widehat{p}_{2,2} & \cdots & \widehat{p}_{n-1,2} & \widehat{p}_{n,2} \\ \vdots & & \ddots & \vdots & \vdots \\ 0 & \cdots & & \widehat{p}_{n-1,n-1} & \widehat{p}_{n,n-1} \end{vmatrix}, \quad n = 1, 2, \ldots, \end{cases} \tag{3.31}$$

and

$$\begin{cases} \widehat{p}_0(x) = 1, \\ \widehat{p}_n(x) = \dfrac{(-1)^{n+1}}{\prod_{k=0}^{n}p_{k,k}} \begin{vmatrix} x & x^2 & \cdots & x^{n-1} & x^n \\ p_{1,1} & p_{2,1} & \cdots & p_{n-1,1} & p_{n,1} \\ 0 & p_{2,2} & \cdots & p_{n-1,2} & p_{n,2} \\ \vdots & & \ddots & \vdots & \vdots \\ 0 & \cdots & & p_{n-1,n-1} & p_{n,n-1} \end{vmatrix}, \quad n = 1, 2, \ldots. \end{cases} \tag{3.32}$$

Proof. Let $\{p_n\}$ be a b. p. s., and let $p_n(x) = \sum_{k=0}^{n} p_{n,k}x^k$ be its decomposition on monomials x^k. We have the sequence (b_n) such that $b_i = p_{i,1}$, and from these we can calculate the coefficients $\widehat{b}_i$ and $\widehat{p}_{i,k}$ by Algorithm 2.1.2. From Theorem 3.4, we get the recurrence relation (3.19), which is equivalent to (3.31) after expanding the determinant with respect to the last column [44].

Vice versa, we suppose that (3.31) is true. Expanding the determinant with respect to the first row, we have $p_n(x) = \sum_{k=0}^{n} p_{n,k}x^k$, where

$$\begin{cases} p_{n,0} = \delta_{n,0}, \\ p_{n,k} = \frac{(-1)^{n-k}}{\prod_{k=1}^{n} \widehat{p}_{i,i}} \begin{vmatrix} \widehat{p}_{k+1,k} & \widehat{p}_{k+2,k} & \cdots & \widehat{p}_{n-1,k} & \widehat{p}_{n,k} \\ \widehat{p}_{k+1,k+1} & \widehat{p}_{k+2,k+1} & \cdots & \widehat{p}_{n-1,k+1} & \widehat{p}_{n,k+1} \\ & \ddots & & & \vdots \\ & & & \widehat{p}_{n-1,n-1} & \widehat{p}_{n,n-1} \end{vmatrix}, & k = 1, \dots, n. \end{cases} \tag{3.33}$$

If we set $P := (p_{n,k})$ and $\widehat{P} = (\widehat{p}_{n,k})$, from (3.33), these matrices are inverse of each other. Then, if we set $\widehat{p}_n(x) = \sum_{k=0}^{n} \widehat{p}_{n,k}x^k$, relation (3.7) holds, and the result follows.

In analogy, we get (3.32). □

Remark 3.9. For a numerical sequence $\{b_n\}_{n\in\mathbb{N}}$ with $b_0 = 0$ and $b_1 \neq 0$, if $\{\widehat{b}_n\}$ and $\{\widehat{p}_{n,k}\}_{n=0,1,\dots,\ k=0,\dots,n}$ are defined in Algorithm 2.1.2, we have the equivalence

$$(3.31) \Leftrightarrow (3.15).$$

We observe explicitly that by (3.15), we can directly obtain (3.31). In fact, if $\{p_n\}$ is such that

$$\exp(x(f(t)) = \sum_{n=0}^{\infty} p_n(x)\frac{t^n}{n!},$$

by substituting $t = \overline{f}(t)$, where $\overline{f}(t)$ is the compositional inverse of $f(t)$ [167], we have

$$\sum_{n=0}^{\infty} p_n(x)\frac{(\overline{f}(t))^n}{n!} = \exp(xt).$$

From (2.2), after some calculations, the previous equality generates an infinite, lower triangular system in the unknowns $p_n(x)$. Solving the first $n + 1$ equations by using Cramer's rule, we get (3.31).

Theorem 3.8 (Second determinantal form). *Let $\{p_n\}$ be a polynomial sequence with $p_0(x) = 1$. It is a b. p. s. if and only if there exists a numerical sequence $(b_i)_i$ with $b_0 = 0$*

and $b_1 \neq 0$ *such that, if* $(\widehat{c}_i)_i$ *are defined in (3.26), it results in*

$$p_0(x) = 1,$$

$$p_{n+1}(x) = \begin{vmatrix} x & -1 & 0 & \cdots & 0 & 0 \\ 0 & x - \binom{1}{0}\widehat{c}_1 & -1 & \cdots & 0 & 0 \\ 0 & -\binom{2}{0}\widehat{c}_2 & x - \binom{2}{1}\widehat{c}_1 & \cdots & 0 & 0 \\ \vdots & \vdots & \vdots & \ddots & \ddots & \vdots \\ 0 & -\binom{n-1}{0}\widehat{c}_{n-1} & -\binom{n-1}{1}\widehat{c}_{n-2} & \cdots & x - \binom{n-1}{n-2}\widehat{c}_1 & -1 \\ 0 & -\binom{n}{0}\widehat{c}_n & -\binom{n}{1}\widehat{c}_{n-1} & \cdots & -\binom{n}{n-2}\widehat{c}_2 & x - \binom{n}{n-1}\widehat{c}_1 \end{vmatrix}. \tag{3.34}$$

Proof. For simplicity, we assume that $b_1 = 1$, and so $\widehat{c}_0 = 1$.

Let $\{p_n\}$ be a b. p. s. From the Theorem 3.5, we get (3.25); this generates an infinite linear system. Then the result follows by considering the first $n + 1$ equations of the linear system in the unknowns $p_i(x)$, $i = 1, \ldots, n + 1$ and by solving it by Cramer's method.

We suppose that (3.34) holds, vice versa. By expanding with respect to the last row, we have (3.25) and, hence, the result. □

Corollary 3.6. *Let* $\{p_n\}_n$ *be a b. p. s. Then* $p_n(x)$ *is the characteristic polynomial of a suitable Hessenberg matrix.*

Proof. It follows, considering that (3.34) is equivalent to

$$p_{n+1}(x) = \det(xI - \widehat{R}),$$

where $\widehat{R} = (\widehat{r}_{i,j})_{i=0,1,\ldots,\ j=0,\ldots,i+1}$ with

$$\widehat{r}_{i,j} = \begin{cases} \binom{i}{j-1}\widehat{c}_{i-j+1}, & 1 \leq j \leq i+1, \\ 0 & \text{otherwise.} \end{cases}$$

□

Remark 3.10. We observe that the matrix $\widehat{R}$ is the production matrix of the matrix $\widehat{P}$.

Remark 3.11. The second recurrence relation and the second determinantal form for a b. p. s. are in [213, Th. 2.5] only as a necessary condition, but they are determined by a more general and complex procedure.

From the previous theorems, we get the following *chain of equivalences*:

$$(3.3) \Leftrightarrow (3.9) \Leftrightarrow (3.15) \Leftrightarrow (3.25) \Leftrightarrow (3.34).$$

Now we can call b.p.s any polynomial sequence, which satisfies one of the above equivalent formulas.

Remark 3.12. We observe that the first determinantal form is an Hessenberg determinant, and it is known that for its calculation the Gaussian elimination is stable [105, pp. 28].

3.5 Relationship with δ-functionals

We shall now consider the relationship with δ-functional connected to Rota et al.'s theory [166, 167].

Definition 3.2 ([167, p. 105], [166, p. 10]). A δ-functional on $\mathcal{P}_n$ is a linear functional such that

$$L(1) = 0, \quad L(x) \neq 0. \tag{3.35}$$

Let $(\widehat{b}_n)_n$ be a numerical sequence with $\widehat{b}_0 = 0$ and $\widehat{b}_1 \neq 0$, and let $\{p_n\}_n$ be the b. p. s. defined by (3.31). We can consider the δ-functional L defined by

$$L(x^i) = \widehat{b}_i, \quad i = 0, \dots, n, \dots. \tag{3.36}$$

Now we define the powers of L as follows [167, p. 101]:

$$L^1 = L, \quad L^k = LL^{k-1}, \quad k \geq 2, \tag{3.37}$$

with

$$L^2(x^i) = \sum_{k=0}^{i} \binom{i}{k} L(x^k) L(x^{i-k}), \quad i = 0, \dots, n. \tag{3.38}$$

We note explicitly that

$$L^2(x^i) = 2\widehat{b}_{i,2}, \quad L^k(x^i) = k!\widehat{b}_{i,k}, \quad k \geq 2,$$

$\widehat{b}_{i,k}$ being defined by Algorithm 2.1.1. We set $L^0 = \varepsilon(0)$, that is, the evaluation functional [167, p. 101].

The product of two linear functionals can be computed using any b. p. s. in place of the sequence x^n.

If $\{p_n\}_n$ is a b. p. s. and $L_1, \dots, L_k$, $k \geq 2$, are linear functionals, then we have [167, p. 103]:

$$(L_1 \cdots L_k)(p_n(x)) = \sum \binom{n}{j_1, \dots, j_k} L_1(p_{j_1}) \cdots L_k(p_{j_k}), \tag{3.39}$$

where the sum is over all k-tuples of nonnegative integers $(j_1, \dots, j_k)$ such that $j_1 + \cdots + j_k = n$.

Then we consider the numerical sequence $(b_i)_i$, conjugate of $(\widehat{b}_i)_i$. Defining the linear functional $\widehat{L}$ as

$$\widehat{L}(x^i) = b_i, \quad i = 0, 1, \dots,$$

we get the following:

Theorem 3.9. *With the previous notations,* $\{p_n\}$ *and* $\{\widehat{p}_n\}$ *are conjugate b. p. s. related to the sequences* (b_i) *and* $(\widehat{b}_i)$ *if and only if*

$$L^k(p_n(x)) = n!\delta_{n,k}, \quad \widehat{L}^k(\widehat{p}_n(x)) = n!\delta_{n,k}, \quad k = 0, \dots, n. \tag{3.40}$$

Proof. We prove the first equality in (3.40). The second one follows from symmetry. It is immediately observable, from (3.39), that $L^k(x^n) = k!\widehat{p}_{n,k}$, and from (3.31) we have the result.

Vice versa, the orthogonality relation (3.40) can be interpreted as a general linear interpolation problem [76, p. 26]. In fact, let us consider the linear functionals

$$L_k = L^k, \quad k = 0, \dots, n;$$

since they are linearly independent, (3.31) allows us to assert that this polynomial $p_n(x)$ is the solution of the linear general interpolation problem

$$L_k(p_n(x)) = n!\delta_{n,k}, \quad k = 0, \dots, n. \tag{3.41}$$

Hence, $p_n(x)$ is given by (3.31). □

Remark 3.13. In [167, pp. 105–106], the orthogonality condition (3.40) characterizes the sequence associated with the δ-functional and the b. p. s.

Likewise, Theorem 3.9 coincides with Theorem 2.3.1 in [166, p. 37] in the b. p. s. case.

Theorem 3.10 (Representation theorem). *If* L *is the* δ*-functional from Definition* 3.2, *then for all* $q_n(x) \in \mathcal{P}_n$, *we have*

$$q_n(x) = \sum_{k=0}^{n} \frac{L^k(q_n(x))}{k!} p_k(x). \tag{3.42}$$

Proof. The result follows from Theorem 3.9. □

We note, explicitly, that (3.42) is the representation of $q_n(x)$ in the basis $\{p_n\}$.

3.6 Relationship with δ-operators

Roman and Rota [167, pp. 129] defined a δ-operator Q on $\mathcal{P}_n$ as follows:

Definition 3.3. A linear operator Q on $\mathcal{P}_n$ is a δ-operator if:
1. it is shift invariant, that is, it commutes with the shift operator E,
2. $Q(1) = 0$ and $Q(x) \neq 0$.

Moreover, a δ-operator Q has the form [93, p. 41]

$$Qy = \sum_{i=1}^{\infty} c_i \frac{y^{(i)}}{i!}, \quad c_i \in \mathbb{R}, \quad c_1 \neq 0, \quad y \in \mathcal{P}. \tag{3.43}$$

This is also the operator defined by Sheffer for A-zero-type set [179, p. 594].

Theorem 3.11 (Associated sequence to δ-operator, [142]). *The following propositions hold:*

1. *For any δ-operator Q, there exists one and only one b. p. s. $\{p_n\}_n$ such that*

$$\begin{aligned} p_0(x) &= 1, \\ Qp_n &= np_{n-1}, \quad n = 1, \ldots. \end{aligned} \tag{3.44}$$

2. *For any b.p.s $\{p_n\}_n$, there exists one and only one δ-operator Q, such that $Qp_n = np_{n-1}$, $n = 1, \ldots$.*

This b.p.s $\{p_n\}_n$ is said to be associated with δ-operator Q.

Proof.

1. Given the δ-operator $Q = \sum_{i=1}^{\infty} c_i \frac{y^{(i)}}{i!}$, $c_1 \neq 0$, we define the numerical sequence $(\widehat{b}_i)_i$ with $\widehat{b}_0 = 0$, $\widehat{b}_i = c_i$, $i = 1, \ldots$, and we consider the b. p. s. $\{p_n\}_n$ as in (3.31). By Theorem 3.3 it holds that

$$\exp(xf(t)) = \sum_{n=0}^{\infty} p_n(x) \frac{t^n}{n!}$$

with $f(t) = \sum_{i=1}^{\infty} b_i \frac{t^i}{i!}$, where (b_i) is the conjugate sequence of $(\widehat{b}_i)$.
Then from the linearity of Q and from relations (2.6) and (2.7), we get

$$\begin{aligned} \sum_{n=0}^{\infty} Qp_n(x)\frac{t^n}{n!} &= Q \sum_{n=0}^{\infty} p_n(x)\frac{t^n}{n!} = Q\exp(xf(t)) \\ &= \left(\widehat{b}_1 f(t) + \frac{\widehat{b}_2}{2} f^2(t) + \cdots + \widehat{b}_n \frac{f^n(t)}{n!} + \cdots\right) \exp(xf(t)) \\ &= \left(\sum_{n=1}^{\infty}\left(\sum_{k=1}^{n} \widehat{b}_k p_{n,k}\right)\frac{t^n}{n!}\right) \exp(xf(t)) \\ &= t\exp(xf(t)) = \sum_{n=0}^{\infty} p_n(x)\frac{t^{n+1}}{n!}. \end{aligned}$$

Then the result follows.

2. If $\{p_n\}_n$ is a b. p. s., we can calculate the sequence $(\widehat{b}_i)_i$ as the conjugate of $(b_i)_i$ with $b_i = p_{i,1}$.
Then the operator

$$Qy = \sum_{i=0}^{\infty} \widehat{b}_i \frac{y^{(i)}}{i!} \tag{3.45}$$

is the only one that satisfies $Qp_n = np_{n-1}$, $n = 1, \ldots$. ☐

Theorem 3.12 (Representation theorem). *Let Q be the δ-operator associated with the b. p. s. $\{p_n\}_n$. Then for any $q_n(x) \in \mathcal{P}_n$, we have*

$$\begin{aligned} q_n(x) &= \sum_{k=0}^{n} \frac{(Q^k q_n(x))_{x=0}}{k!} p_k(x) = \sum_{k=0}^{n} \left(\sum_{i=k}^{n} \widehat{b}_{i,k} (q_n^{(i)}(x))_{x=0} \right) \frac{p_k(x)}{k!} \\ &= \sum_{k=0}^{n} \frac{q_n^{(k)}(0)}{k!} z_k(x) = \sum_{k=0}^{n} \frac{q_n^{(k)}(0)}{k!} x^k, \end{aligned} \tag{3.46}$$

where

$$z_k(x) = \sum_{i=0}^{k} \widehat{b}_{i,k} p_i(x).$$

Proof. We observe that if $Qy = \sum_{i=0}^{\infty} \widehat{b}_i \frac{y^{(i)}}{i!}$, then defining the linear functional

$$L(x^i) = (Qx^i)_{x=0},$$

the result is

$$L(x^i) = \widehat{b}_i, \quad L^k(x^i) = \widehat{b}_{ik}.$$

Thus, the result follows by Theorems 3.10 and 3.7. Moreover, the last identity follows from

$$z_k(x) = \sum_{i=0}^{k} \widehat{b}_{k,i} p_i(x) = x^k, \quad k = 0, \dots, n.$$

□

Identity (3.46) is the Taylor polynomial in the basis $\{p_k\}_{k=0,\dots,n}$, and we call it the *Taylor polynomial relative to the δ-operator Q*.

We note that Theorem 3.12, in the first equality, coincides with Aitken's theorem [11] in the particular case of the choice $\theta_1 = \theta_2 = \cdots = \theta_n = Q$. That is, does the Aitken's theorem discover or rediscover the b. p. s.?

Theorem 3.13 (Functional equation). *The b. p. s. $\{p_n\}_n$ satisfies the functional equation [66, Th. 4.10]*

$$\sum_{i=0}^{n-1} i! \widehat{p}_{ni} Q^{n-i} y(x) + n! \widehat{p}_{nn} y(x) = n! x^n. \tag{3.47}$$

Proof. From (3.44), $Qp_n = np_{n-1}$, so we have

$$Q^k p_n(x) = \frac{k!}{n!} p_{n-k}(x), \quad k = 1, \dots, n-1.$$

Replacing $p_k(x)$ with the previous identity, in the first recurrence formula

$$p_n(x) = \frac{1}{\widehat{p}_{nn}} \left(x^n - \sum_{k=0}^{n-1} \widehat{p}_{nk} p_k(x) \right),$$

we have the result. □

Proposition 3.6. *The b. p. s.* $\{p_n\}_n$ *satisfies the differential equation*

$$\forall n > 1, \quad \sum_{k=1}^{n-1} K_k p_n^{(k)}(x) + n!\widehat{p}_{nn} p_n(x) = n!x^n, \tag{3.48}$$

where

$$K_j = \sum_{i=1}^{n-j} i!\widehat{p}_{n,i}\widehat{p}_{j,n-i}. \tag{3.49}$$

Proof. Given that

$$Qy = \sum_{i=1}^{\infty} \widehat{b}_i \frac{y^{(i)}}{i!}, \quad Q^k y = \sum_{i=k}^{\infty} \widehat{b}_{i,k} \frac{y^{(i)}}{i!}$$

with $\widehat{b}_{i,k} := \widehat{p}_{i,k}$, and $p_n(x) \in \mathcal{P}_n$, the result follows from Theorem 3.13 after easy calculations. ☐

Corollary 3.7 (Fourth recurrence relation). *For a b. p. s.* $\{p_n\}_n$, *the following recurrence relation holds:*

$$\forall n > 1: \quad p_n(x) = \frac{1}{\widehat{p}_{nn}}\left(x^n - \sum_{i=1}^{n-1} K_i p_n^{(i)}(x)\right). \tag{3.50}$$

Remark 3.14. Other differential equations for b. p. s. have been determined by H. Young and Y. Yang [215].

3.7 Connection constants

The problem of computing the connection constants between two b. p. s. is very important.

Given two b. p. s. $\{p_n\}_n$ and $\{q_n\}_n$, we want to determine the coefficients $c_{n,k}$, such that

$$q_n(x) = \sum_{k=0}^{n} c_{n,k} p_n(x).$$

In connection to this aim, we have the following:

Theorem 3.14 (Connection constants). *Let* $P_n(x) = [p_0(x), p_1(x), \dots, p_n(x)]^T$ *and* $Q_n(x) = [q_0, q_1, \dots, q_n(x)]^T$ *be binomial vectors such that:*

$$P_n(x) = AX_n(x)$$

with

$$A = \begin{cases} a_{i,j}, & i = 0, \dots, n, \quad j = 0, \dots, i, \\ 0 & \textit{otherwise}, \end{cases}$$

and

$$Q_n(x) = BX_n(x)$$

with

$$B = \begin{cases} b_{i,j}, & i = 0, \dots, n, \quad j = 0, \dots, i, \\ 0 & \textit{otherwise}. \end{cases}$$

Then

$$P_n(x) = CQ_n(x), \tag{3.51}$$

where

$$C = \begin{cases} c_{i,j}, & i = 0, \dots, n, \quad j = 0, \dots, i, \\ 0 & \textit{otherwise}, \end{cases}$$

with

$$c_{i,j} = \sum_{k=j}^{i} a_{i,k} \widehat{b}_{k,j}, \tag{3.52}$$

$\widehat{b}_{i,j}$ *being the elements of the matrix* B^{-1}.

Proof. We have $X_n(x) = B^{-1}Q_n(x)$. Therefore, substituting, we get

$$P_n(x) = AB^{-1}Q_n(x).$$

So the theorem is proved by setting $C = AB^{-1}$. □

Corollary 3.8. *In the hypothesis of the previous theorem, the result is*

$$\forall n \in \mathbb{N}, \quad p_n(x) = \sum_{k=0}^{n} c_{n,k} q_k(x). \tag{3.53}$$

This relation means that the b. p. s. $\{p_n\}_n$ is the umbral composition of the polynomial sequence $\{r_n\}_n$ with $r_n(x) := \sum_{k=0}^{n} c_{n,k} x^k$ and the b. p. s. $\{q_n\}_n$.

3.8 Summary

Let $(b_i)_i$ be a numerical sequence with $b_0 = 0$ and $b_1 \neq 0$. From this sequence, we have defined the matrix $P := (p_{i,j})$, called a binomial-type matrix. Then we have considered the numerical sequence $(\widehat{b}_i)_i$ and the matrix $\widehat{P} := (\widehat{p}_{i,j})$. The matrix $\widehat{P}$ is also of binomial-type, and it results in $P^{-1} = \widehat{P}$. The matrices P and $\widehat{P}$ generate two polynomial sequences, denoted—respectively—$\{p_n\}$ and $\{\widehat{p}_n\}$, called conjugate, because they are the inverses of each other with respect to the umbral composition, that is, $p_n(\widehat{p}_n(x)) = \widehat{p}_n(p_n(x)) = x^n$. These polynomial sequences are regarded as of binomial type. They satisfy:

1. $p_0(x) = 1, p_n(0) = 0, \forall n > 0$, and

$$p_n'(x) = \sum_{k=0}^{n-1} \binom{n}{k} b_{n-k} p_k(x);$$

2. the binomial identity

$$p_n(x + y) = \sum_{k=0}^{n} \binom{n}{k} p_k(x) p_{n-k}(y);$$

3. some recurrence relations and relative determinantal forms;
4. differential relations.

Finally, the generating function and relationship with δ-functional and δ-operators are determined.

4 Applications to linear interpolation and operators approximation theory

Abstract: The polynomial sequences of binomial type have applications in approximation theory of regular functions. In what follows, we focus on the linear interpolation and uniform approximation.

4.1 Linear interpolation

The representation Theorem 3.12 for polynomials can be extended to some classes of real functions.

Let X be the linear space of real functions defined in the interval $I := [a, b]$, continuous and with continuous derivatives of all necessary orders. Let $\mathcal{P}_n \subset X$, $n \in \mathbb{N}$, be the set of polynomials of degree less than or equal to n. Let Q be a δ-operator on $\mathcal{P}_n$ and $\mathrm{XQ} = \{f \in \mathrm{X} \mid Q^i f \in \mathrm{X},\ i = 0, 1, \dots, n\ \forall n \in \mathbb{N}\}$.

If $\{p_n\}_{n\in\mathbb{N}}$ is the associated sequence to Q, then we have

Theorem 4.1 (Binomial interpolation, [14, 67]). *For any $f \in \mathrm{XQ}$, we consider the polynomial*

$$P_n[f](x) = \sum_{k=0}^{n} (Q^k f)_{x=0} \frac{p_k(x)}{k!}. \tag{4.1}$$

Then

$$(Q^k f)_{x=0} = Q^k (P_n[f])_{x=0}, \quad k = 0, \dots, n. \tag{4.2}$$

Moreover,

$$f(x) = P_n[f](x) + R_n[f](x), \quad x \in [a, b], \tag{4.3}$$

where

$$R_n[f](x) = \int_a^b K_n(x, t) f^{(n+1)}(t)dt, \tag{4.4}$$

and

$$K_n(x, t) = \frac{1}{n!}\left[(x - t)_+^n - \sum_{k=0}^{n} Q^k([(x - t)_+^n])_{x=0} \frac{p_k(x)}{k!}\right]. \tag{4.5}$$

https://doi.org/10.1515/9783110652925-004

Proof. The interpolant conditions (4.2) follow from properties of associated sequence $\{p_n\}$. Moreover, we get (4.3) by observing that the linear functional

$$R_n[q](x) = q(x) - \sum_{k=0}^{n} (Q^k q)_{x=0} \frac{p_k(x)}{k!}$$

vanishes for any $q \in \mathcal{P}_n$, and, hence, we can apply Peano's lemma [76, p. 70]. □

The polynomial (4.1) is called the *binomial interpolant polynomial of f* relative to δ-operator Q and related associated sequence $\{p_n\}$; this can be considered the Taylor polynomial of a regular function with respect to the δ-operator Q. In fact, if $Q = D$, we have $\{p_n\} = \{x^n\}$, and the polynomial (4.1) is the classical Taylor polynomial.

We also note that the polynomial (4.1) for $f \in \mathcal{P}_n$ coincides with the general formula of polynomial interpolation of Aitken [11] in the particular case $\theta_1 = \cdots = \theta_n = Q$.

Hence, the notion of b. p. s. is inherent in the Aitken theorem.

4.2 Approximation operators of binomial type

The application of binomial sequences to linear operators approximation theory has been fully debated (see [10, 111, 133, 134, 141, 152, 154, 172, 182–184, 189] and references therein). In fact, the b. p. s. allows us to generalize the classical Bernstein operator [13, 25, 36, 129]. One of the first possibilities to obtain such operators was given by Popoviciu [154] in 1932. The author introduced the linear operator $T_n : C[0,1] \to C[0,1]$ defined by

$$(T_n f)(x) = \frac{1}{p_n(1)} \sum_{k=0}^{n} \binom{n}{k} p_k(x) p_{n-k}(1-x) f\left(\frac{k}{n}\right), \tag{4.6}$$

where $\{p_k\}_k$ is a b. p. s., which satisfies the conditions

$$p_n(1) \neq 0, \quad p_n(x) \geq 0 \quad \text{in} \quad [0,1], \quad n = 0, 1, \ldots.$$

Remark 4.1. For the b. p. s. $p_n(x) = x^n$, the operator (4.6) is the well-known Bernstein polynomial operator.

Popoviciu also gave the convergence property of this operator, but he has not emphasized these ideas in other publications. So other authors [172, 183] have considered the same operator as being of *binomial-type* or *Bernstein–Sheffer operator of degree n* [172]. This operator, denoted with $L_n : C[0,1] \to C[0,1]$, is defined by

$$(L_n \phi)(x) = \frac{1}{p_n(1)} \sum_{k=0}^{n} \binom{n}{k} p_k(x) p_{n-k}(1-x) \phi\left(\frac{k}{n}\right), \tag{4.7}$$

where $\{p_k\}_k$ is a b. p. s., which satisfies the condition $p_n(1) \neq 0$.

Then we get:

Theorem 4.2. *The operator L_n defined in 4.7 is a positive operator on $C[0,1]$ for all $n \geq 0$ if and only if $b_i \geq 0$ for $i = 1, 2, \ldots$, where (b_i) is the numerical sequence that generates the b. p. s. $\{p_n\}_n$.*

Proof. The proof follows from Proposition 3.1. □

In [172], the following results are proved:

Theorem 4.3. *Assume that $\{p_n\}_n$ is a b. p. s. that satisfies the hypothesis of Theorem 4.2. Then:*

1. *L_n is an isomorphism of $\mathcal{P}_n$, reserving the degree, that is, $L_n(p) \in \mathcal{P}_k$ if $p \in \mathcal{P}_k$ for $0 \leq k \leq n$.*
2. *Putting $e_i = x^i$, $i = 0, 1, \ldots$, one has*

$$L_n e_0 = e_0, \quad L_n e_1 = e_1,$$

and

$$L_n e_2 = e_2 + a_n(e_1 - e_2),$$

where

$$a_n = 1/n + ((n-1)/n)(r_{n-2}/p_n), p_n = p_n(1), \quad r_n = r_n(1),$$

and the sequence $\{r_n\}_n$ is generated by

$$f''(t)\exp(xf(t)) = \sum_{n\geq 0} r_n(x)\frac{t^n}{n!} \tag{4.8}$$

with $f(t) = \sum_{i=0}^{\infty} b_i \frac{t^i}{i!}$.

Theorem 4.4. *Assume that L_n satisfies the condition of Theorems 4.2, 4.3. Then:*

1. *$L_n\phi$ converges uniformly to ϕ, $\forall \phi \in C[0,1]$, when n tends to infinity if and only if*

$$\lim_{n\to\infty} \frac{r_{n-2}}{p_n} = 0. \tag{4.9}$$

2. *More specifically, if $r_{n-2}/p_n = O(1/n)$, then there exists an integer $k \geq 1$ such that for any δ-series $f(t) \in \mathcal{P}_k$, we have*

$$\|\phi - L_n\phi\|_\infty \leq \left(1 + \frac{1}{2}\sqrt{k}\right)\omega\left(\phi, \frac{1}{\sqrt{n}}\right), \tag{4.10}$$

where ω is the modulus of continuity of ϕ [13].

It is known that for the classic Bernstein operator, relation (4.10) holds with $k = 1$. Therefore, we can say that the Bernstein operator is the best positive binomial–Sheffer operator associated with the function $f(t) \in \mathcal{P}_k$, $k \geq 1$.

A Voronovskaya-type theorem [202] is also true.

Theorem 4.5. *If $\phi \in C^2[0,1]$ in the hypothesis of Theorems 4.2, 4.3, and 4.4, the following statement holds:*

$$\lim_{n\to\infty} n[L_n\phi - \phi] = \frac{1}{2}x(x-1)\phi''(x), \quad x \in I, \tag{4.11}$$

and the convergence is uniform.

Proof. We observe that

$$\phi\left(\frac{k}{n}\right) = \phi(x) + \left(\frac{k}{n} - x\right)\phi'(x) + \left(\frac{k}{n} - x\right)^2 \left[\frac{1}{2}\phi''(x) + h\left(\frac{k}{n} - x\right)\right],$$

where $h(y) := h_n(y)$ is bounded for all y and converges to zero with y. The result follows from the above theorems. □

Theorem 4.5 shows that the convergence cannot be faster than $\frac{1}{n}$, even for smooth functions.

To improve the convergence, an extrapolation technique can be applied.

In fact, we get:

Theorem 4.6. *If $\phi(x) \in C^2[a,b]$ satisfies the conditions of Theorem 4.4, then the linear operator*

$$T_{2n}[\phi] := 2L_{2n}[\phi](x) - L_n[\phi](x) \tag{4.12}$$

converges uniformly to ϕ in $[0,1]$, and

$$T_{2n}[\phi](x) = \phi(x) + O\left(\frac{1}{n^2}\right) \quad \forall x \in [0,1], \quad n \to \infty. \tag{4.13}$$

Proof. The proof follows with the technique of [45, Th. 3.1]. □

Ismail [111] presents a more general class of linear positive operators, called the *exponential-type*, connected to the b. p. s. $\{p_n\}_n$ with $p_n(x) \geq 0$, $\forall x \geq 0$ and $n = 0, 1, \ldots$. He proves a one-to-one correspondence between positive b. p. s. and exponential-operators.

The polynomial sequences of binomial type have also been used in approximation theory in semiinfinite interval. In fact, in [189], it is considered the operator sequence:

$$\tilde{T}_n[f](x) = \exp(-2(\sqrt{2}-1)nx) \sum_{k=0}^{\infty} \frac{\theta_k(2\sqrt{2}nx)}{4^k k!} f\left(\frac{k}{n}\right), \tag{4.14}$$

where $\theta_k(x)$ is the b. p. s., called reverse Bessel polynomials [33, 166] and defined by

$$\theta_k(x) = x^k y\left(\frac{1}{x}\right) = \sum_{j=1}^{k} \frac{(2k-j-1)!}{(j-1)!(k-j)!2^{2k-j}} x^j$$

with $y(x)$ the Bessel polynomials [14, 33].

The operator sequence (4.14), on suitable hypothesis, converges in the semiinfinite interval $[0, +\infty[$ for functions of exponential type. Moreover, the following quantitative error estimate holds:

$$|\tilde{T}_n[f](x) - f(x)| \leq \left(1 + \sqrt{\frac{3}{2}x}\right)\omega\left(f, \frac{1}{\sqrt{n}}\right),$$

given $\omega(f, \frac{1}{\sqrt{n}})$, the modulus of continuity of f.

4.3 Summary

Let Q be a δ-operator, and let $\{p_n\}$ be the associated b. p. s. It is possible to represent any polynomial $P_n \in \mathcal{P}_n$ by the b. p. s. $\{p_n\}_n$ and the δ-operator Q. For real functions in the set XQ, we can consider an interpolating polynomial. For operators approximation theory, the b. p. s. allows us to generalize the classical Bernstein operator. Under suitable hypothesis, an extrapolation is also possible, even though there is a lack of complete asymptotic expansion in literature.

Finally, the polynomial sequences of binomial type have been used in approximating functions in semiinfinite interval by generalized Szasz operators [189].

5 Examples

Abstract: In this chapter, we will discuss some important examples of b. p. s. without any pretence of doing a complete treatise on b. p. s.

We will follow this order:
1. Numerical sequences: $b := (b_n)_{n\in\mathbb{N}}$ and $\hat{b} := (\hat{b}_n)_{n\in\mathbb{N}}$;
2. Exponential Riordan matrices associated: P and $\hat{P}$;
3. Polynomial sequences $\{p_n\}_n$ and $\{\hat{p}_n\}_n$;
4. Generating function, binomial identity, recurrence and differential equations, determinantal forms;
5. δ-functional and δ-operator associated, functional equation;
6. Applications to interpolation and approximation.

Naturally the binomial identity, the functional equation, et cetera, will be repeated only if they are explicit or significant.

5.1 The sequence $\{x^n\}_{n\in\mathbb{N}}$

Let us consider the numerical sequence

$$b_0 = 0, \quad b_1 = 1, \quad b_i = 0, \quad i = 2, \ldots, \tag{5.1}$$

then from (2.5), we get

$$\hat{b}_0 = 0, \quad \hat{b}_1 = 1, \quad \hat{b}_i = 0, \quad i = 2, \ldots. \tag{5.2}$$

From Algorithms 2.1.1 and 2.1.2, we get

$$b_{n,k} = \hat{b}_{n,k} = \delta_{n,k}, \quad k = 0, 1, \ldots n \geq k,$$

and, therefore,

$$p_{n,k} = \hat{p}_{n,k} = \delta_{n,k}.$$

The associated matrices are

$$P = \hat{P} = I.$$

From (3.2), we get the polynomial sequence

$$p_n(x) = \hat{p}_n(x) = x^n. \tag{5.3}$$

https://doi.org/10.1515/9783110652925-005

The binomial identity is the well-known identity

$$(x+y)^n = \sum_{k=0}^{n} \binom{n}{k} x^k y^{n-k}.$$

The associated operator is

$$Qy = \widehat{Q}y = y'(x). \tag{5.4}$$

Hence, $Q = D$, and therefore the representation theorem is

$$q(x) = \sum_{k=0}^{n} (q^{(k)}(x))_{x=0} \frac{x^k}{k!} \quad \forall q(x) \in \mathcal{P}_n, \tag{5.5}$$

that is, the Taylor expansion with initial point $x = 0$.

Consequently, the interpolation polynomial for $f \in C^n[0,1]$ is the well-known Taylor polynomial, whereas the operator approximation (4.7) is the classic Bernstein operator [25].

5.2 Lower factorial and exponential polynomials

Let us consider the sequence $\widehat{b} := (\widehat{b}_i)$ with

$$\widehat{b}_0 = 0, \quad \widehat{b}_i = 1, \quad i = 1, 2, \ldots. \tag{5.6}$$

From the sequence $\widehat{b}$, we can calculate the conjugate sequence $b := (b_i)$, applying (2.7) as

$$b_0 = 0, \quad b_1 = 1, \quad b_n = (-1)^{n-1}(n-1)!, \quad n = 2, \ldots. \tag{5.7}$$

Remark 5.1. The associated formal power series to $\widehat{b}$ and b are, respectively,

$$\overline{f}(t) = \sum_{i=1}^{\infty} \frac{t^i}{i!} = e^t - 1 \tag{5.8}$$

and

$$f(t) = \sum_{i=1}^{\infty} (-1)^{i-1} \frac{t^i}{i} = \log(1+t). \tag{5.9}$$

From Algorithm 2.1.1, we get

$$\begin{cases} \widehat{p}_{n,0} = \delta_{n,0}, \\ \widehat{p}_{n,1} = 1, \quad n = 1, 2, \ldots, \\ \widehat{p}_{n,k} = \frac{1}{k} \sum_{i=1}^{n-k+1} \binom{n}{i} \widehat{p}_{n-i,k-1}, \quad k = 2, \ldots, \quad n \geq k, \end{cases} \tag{5.10}$$

and

$$\begin{cases} p_{n,0} = \delta_{n,0}, \\ p_{n,1} = (-1)^{n-1}(n-1)!, \quad n = 1, 2, \ldots, \\ p_{n,k} = \frac{1}{k}\sum_{i=1}^{n-k+1}\binom{n}{i}(-1)^{i-1}(i-1)!p_{n-i,k-1}, \quad k = 2, \ldots, \quad n \geq k. \end{cases} \tag{5.11}$$

Remark 5.2. We observe that $\widehat{p}_{n,k} \geq 0$, $k = 0, \ldots, n$, $n = 0, 1, \ldots$, whereas, by induction, we can prove that $sign(p_{n,k}) = (-1)^{n+k}$ and $p_{n,n} = 1$.

For example, for $n = 5$, the matrices P and $\widehat{P}$ are:

$$\widehat{P} = \begin{pmatrix} 1 & 0 & 0 & 0 & 0 & 0 \\ 0 & 1 & 0 & 0 & 0 & 0 \\ 0 & 1 & 1 & 0 & 0 & 0 \\ 0 & 1 & 3 & 1 & 0 & 0 \\ 0 & 1 & 7 & 6 & 1 & 0 \\ 0 & 1 & 15 & 25 & 10 & 1 \end{pmatrix}, \quad P = \begin{pmatrix} 1 & 0 & 0 & 0 & 0 & 0 \\ 0 & 1 & 0 & 0 & 0 & 0 \\ 0 & -1 & 1 & 0 & 0 & 0 \\ 0 & 2 & -3 & 1 & 0 & 0 \\ 0 & -6 & 11 & -6 & 1 & 0 \\ 0 & 24 & -50 & 35 & -10 & 1 \end{pmatrix}.$$

From (3.2), we get the polynomial sequences $\{p_n\}_n$ and $\{\widehat{p}_n\}_n$:

$$p_n(x) = \sum_{i=0}^{n} p_{n,i}x^i, \tag{5.12}$$

$$\widehat{p}_n(x) = \sum_{i=0}^{n} \widehat{p}_{n,i}x^i. \tag{5.13}$$

We easily get the polynomial sequences. For example, the first six polynomials are

$$\begin{aligned} &\widehat{p}_0(x) = 1, && p_0(x) = 1, \\ &\widehat{p}_1(x) = x, && p_1(x) = x, \\ &\widehat{p}_2(x) = x^2 + x, && p_2(x) = x^2 - x, \\ &\widehat{p}_3(x) = x^3 + 3x^2 + x, && p_3(x) = x^3 - 3x^2 + 2x, \\ &\widehat{p}_4(x) = x^4 + 6x^3 + 7x^2 + x, && p_4(x) = x^4 - 6x^3 + 11x^2 - 6x, \\ &\widehat{p}_5(x) = x^5 + 10x^4 + 25x^3 + 15x^2 + x, && p_5(x) = x^5 - 10x^4 + 35x^3 - 50x^2 + 24x. \end{aligned}$$

Their plots are shown in Figures 5.1 and 5.2.

We consider the linear functionals L and $\widehat{L}$ defined by

$$L(x^i) = \widehat{b}_i, \quad \widehat{L}(x^i) = b_i, \quad i = 0, 1, \ldots.$$

Then we have the following:

Theorem 5.1. *The following statement is true:*

$$L^k(p_n(x)) = p_n(k), \quad k = 1, \ldots, n. \tag{5.14}$$

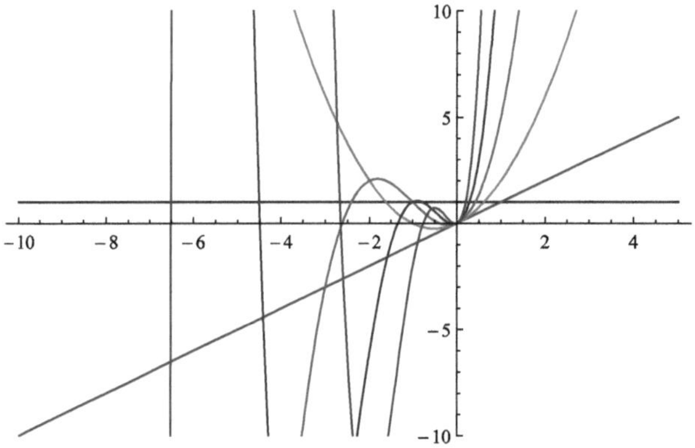

Figure 5.1: Exponential polynomials.

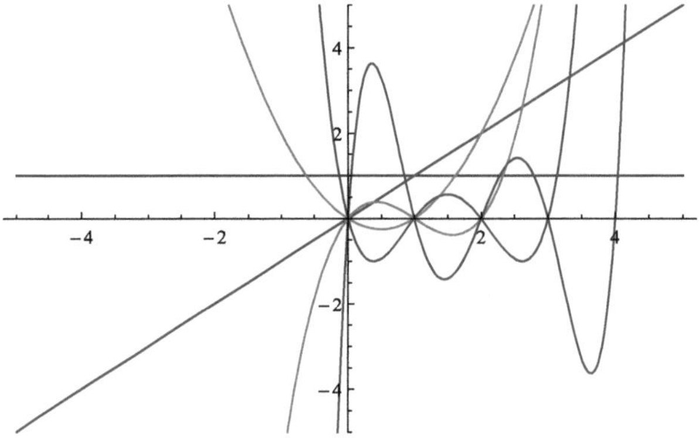

Figure 5.2: Lower factorial polynomials.

Proof. We proceed by mathematical induction on k. For $k = 1$, we have

$$L(p_n) = L\left(\sum_{k=0}^{n} p_{n,k}x^k\right) = \sum_{k=0}^{n} p_{n,k}L(x^k) = \sum_{k=0}^{n} p_{n,k}\widehat{b}_k = \sum_{k=0}^{n} p_{n,k} = p_n(1).$$

Moreover, for $k = 2, \ldots, n$, we get

$$\begin{aligned} p_n(k) = p_n(\underbrace{1 + \cdots + 1}_{k\text{-terms}}) &= \sum \binom{n}{\nu_1, \ldots, \nu_k} p_{\nu_1}(1) \ldots p_{\nu_k}(1) \\ &= \sum \binom{n}{\nu_1, \ldots, \nu_k} L(p_{\nu_1}(x)) \ldots L(p_{\nu_k}(x)) = L^k(p_n(x)). \end{aligned}$$

□

Remark 5.3. For the linear functional L and from the first determinantal form of $\{p_n\}$, the following relation holds:

$$L^k(p_n(x)) = n!\delta_{n,k}.$$

Theorem 5.2 ([66, Th. 5.3]). *For the b. p. s. $\{p_n\}_n$, we have*

$$p_n(x) = x(x-1)\cdots(x-n+1). \tag{5.15}$$

Proof. From Theorem 5.1 and Remark 5.3, $p_n(x)$ has zeros $x = 0, 1, 2, \ldots, n-1$. Hence, the result follows. □

The polynomials $\{p_n\}_n$, from Theorem 5.2, coincide with the polynomials called in literature *lower factorial* or *falling polynomials* [166, p. 56]. They are denoted with $(x)_n$, that is,

$$p_n(x) = x(x-1)\cdots(x-n+1) := (x)_n. \tag{5.16}$$

Then we get from (5.12)

$$(x)_n = \sum_{k=0}^{n} p_{n,k} x^k, \tag{5.17}$$

but in [113, p. 142], it is written

$$(x)_n = \sum_{k=1}^{n} S_n^k x^k,$$

where S_n^k are the Stirling numbers of first kind.

Hence, we deduce that the numbers $p_{n,k}$ coincide with the numbers S_n^k. Following instead the notation in Riordan [164, p. 90], the Stirling numbers of first kind are denoted as $s(n,k)$; therefore, we write

$$(x)_n = \sum_{k=0}^{n} s(n,k)x^k = \sum_{k=0}^{n} p_{n,k} x^k;$$

the same is also according to Roman [166, p. 57].

From (5.16), the matrices P and $\widehat{P}$ being inverse of each other, we get

$$x^n = \sum_{k=0}^{n} \widehat{p}_{n,k} (x)_k \tag{5.18}$$

and comparing with [113, p. 168], we deduce that the numbers $\widehat{p}_{n,k}$ are known in the literature as *Stirling's numbers of second kind* and denoted in [164, 166, 186] by $S(n,k)$. Therefore, we obtain

$$x^n = \sum_{k=0}^{n} \widehat{p}_{n,k} (x)_k = \sum_{k=0}^{n} S(n,k)(x)_k. \tag{5.19}$$

Moreover, we also get

$$\widehat{p}_n(x) = \sum_{k=0}^{n} S(n,k) x^k. \tag{5.20}$$

In this form, the sequence $\{\widehat{p}_n\}$ is known in literature as *exponential polynomial sequence* [22, 29, 66, 164, 166, 212] and denoted by $\phi_n(x)$, that is,

$$\phi_n(x) = \widehat{p}_n(x) = \sum_{k=0}^{n} \widehat{p}_{n,k}x^k = \sum_{k=0}^{n} S(n,k)x^k. \tag{5.21}$$

In [29], some historical notes and applications of exponential polynomial sequence are given. There is a wide literature on this topic [29, 99, 101] and references therein.

The recurrence formulas (5.10) and (5.11) are not known in literature, and they seem to be useful for the calculation of Stirling's numbers. Moreover, directly from Remark 5.2, it follows that the Stirling's numbers of second kind are positive, whereas the Stirlng's numbers of first kind are in alternating signs. We also note that the generating functions of Stirling's numbers are

$$\frac{1}{k!}(\exp(t)-1)^k, \quad \frac{1}{k!}(\log(1+t))^k, \quad \forall k \geq 0,$$

that is,

$$\frac{1}{k!}(\exp(t)-1)^k = \sum_{n=k}^{\infty} S(n,k)\frac{t^n}{n!}, \tag{5.22}$$

$$\frac{1}{k!}(\log(1+t))^k = \sum_{n=k}^{\infty} s(n,k)\frac{t^n}{n!}. \tag{5.23}$$

From (5.22), it follows that $S(n,k) > 0$, $\forall n, k \geq 0$, whereas from (5.23), we get that $s(n,k)$ have alternate signs.

Now we consider the determinantal forms and recurrence and differential relations.

Theorem 5.3 (Determinantal form). *From (3.31), the b. p. s. associated with the sequence b is*

$$p_0(x) = 1, \quad p_n(x) = (-1)^{n+1} \begin{vmatrix} x & x^2 & \cdots & x^{n-1} & x^n \\ 1 & 1 & \cdots & 1 & 1 \\ 0 & 1 & \cdots & S(n-1,2) & S(n,2) \\ \vdots & & \ddots & & \vdots \\ 0 & \cdots & & 1 & S(n,n-1) \end{vmatrix}, \tag{5.24}$$

whereas the b. p. s. associated with the sequence $\widehat{b}$ is

$$\widehat{p}_0(x) = 1, \quad \widehat{p}_n(x) = (-1)^{n+1} \begin{vmatrix} x & x^2 & \cdots & x^{n-1} & x^n \\ 1 & -1 & \cdots & (-1)^{n-2}(n-2)! & (-1)^{n-1}(n-1)! \\ 0 & 1 & \cdots & s(n-1,2) & s(n,2) \\ \vdots & & \ddots & & \vdots \\ 0 & \cdots & & 1 & s(n,n-1) \end{vmatrix}.$$

Proposition 5.1. *The following identity holds:*

$$(-1)^{n+1}\begin{vmatrix} x & x^2 & \cdots & x^{n-1} & x^n \\ 1 & 1 & \cdots & 1 & 1 \\ 0 & 1 & \cdots & S(n-1,2) & S(n,2) \\ \vdots & & \ddots & & \vdots \\ 0 & \dots & & 1 & S(n,n-1) \end{vmatrix} = x(x-1)\cdots(x-n+1), \quad \forall x \in \mathbb{R}. \tag{5.25}$$

Proof. It follows from (5.15) and the determinantal form. □

Theorem 5.4 (Recurrence and differential relations). *For the lower factorial and the exponential polynomials, we have*

$$((x)_n)' = \sum_{i=1}^{n} \binom{n}{i} (-1)^{i-1}(i-1)!(x)_{n-i}, \tag{5.26}$$

$$\phi_n'(x) = \sum_{i=1}^{n} \binom{n}{i} \phi_{n-i}(x), \tag{5.27}$$

$$(x)_n = x^n - \sum_{k=0}^{n-1} S(n,k)(x)_k, \tag{5.28}$$

$$\phi_n(x) = x^n - \sum_{k=0}^{n-1} s(n,k)\phi_k(x), \tag{5.29}$$

$$(x)_{n+1} = -(n\widehat{c}_1 - x)(x)_n - \sum_{k=0}^{n-2} \binom{n}{k} \widehat{c}_{n-k}(x)_{k+1}, \tag{5.30}$$

$$\textit{with} \quad \sum_{k=0}^{n} \binom{n}{k} \widehat{c}_k(-1)^{n-k}(n-k)! = \begin{cases} 1, & n = 0, \\ 0, & n \geq 1, \end{cases}$$

$$\phi_{n+1}(x) = -(n\widehat{d}_1 - x)\phi_n(x) - \sum_{k=0}^{n-2} \binom{n}{k} \widehat{d}_{n-k}\phi_{k+1}(x), \tag{5.31}$$

$$\textit{with} \quad \sum_{k=0}^{n} \binom{n}{k} \widehat{d}_k = \begin{cases} 1, & n = 0, \\ 0, & n \geq 1, \end{cases}$$

$$(x)_{n+1} = x\sum_{k=0}^{n} \binom{n}{k} (-1)^{n-k}(n-k)!(x)_k, \tag{5.32}$$

$$\phi_{n+1}(x) = x\sum_{k=0}^{n} \binom{n}{k} \phi_k(x). \tag{5.33}$$

Proof. The proof follows from Theorems 3.1, 3.4, 3.5, 3.6 and previous results. □

Remark 5.4. Relations (5.27) and (5.33) are also in [29], but they are determined by different methods.

For the zeros of the exponential polynomials $\phi_n(x)$, the following important results hold:

Proposition 5.2 ([29, 101]). *All zeros of $\phi_n(x)$, $\forall n \in \mathbb{N}$, are real, simple, and nonpositive.*

Proposition 5.3. *The exponential polynomials have as rational zeros only $x = 0$ and at the most $x = -1$.*

Proof. It follows easily from the known properties of polynomials with integer coefficients [52]. □

Important relations between exponential polynomials and Bernoulli numbers are known (see, e. g., [29, (2.25), (4.1)]:

$$\int_0^x \phi_n(t)dt = \frac{1}{n+1}\sum_{k=1}^{n+1}\binom{n+1}{k} B_{n-k+1}\phi_k(x),$$

$$\int_0^\infty \phi_n(-x)\phi_m(-x)\frac{\exp(-2x)}{x}dx = (-1)^{n-1}\frac{2^{n+m}-1}{n+m}B_{n+m},$$

where B_k are the Bernoulli numbers [113, p. 234].

The binomial identity can be written as

$$(x+y)_n = \sum_{k=0}^{n}\binom{n}{k}(x)_k(y)_{n-k}, \tag{5.34}$$

$$\phi_n(x+y) = \sum_{k=0}^{n}\binom{n}{k}\phi_k(x)\phi_{n-k}(y). \tag{5.35}$$

We note that for $y = -x$, we get the discrete orthogonality

$$\sum_{k=0}^{n}\binom{n}{k}\phi_k(x)\phi_{n-k}(-x) = 0,$$

$$\sum_{k=0}^{n}\binom{n}{k}(x)_k(-x)_{n-k} = 0.$$

Remark 5.5 (Associated δ-operator). The δ-operator associated with b. p. s. $\{p_n\}$ is

$$Qy(x) = \sum_{i=1}^{\infty}\frac{y^{(i)}(x)}{i!} = y(x+1) - y(x) := \Delta y(x),$$

whereas the δ-operator associated with the b. p. s. $\{\widehat{p}_n\}$ is

$$\widehat{Q}y(x) = \sum_{i=1}^{\infty}(-1)^{i-1}\frac{y^{(i)}(x)}{i}.$$

Theorem 5.5 (Representation theorem). *For any polynomial $q_n(x)$, we have*

$$q_n(x) = \sum_{k=0}^{n}(\Delta^k q_n(x))_{x=0}\frac{(x)_k}{k!}$$

or

$$q_n(x) = \sum_{k=0}^{n} (\widehat{Q}^k q_n(x))_{x=0} \frac{\phi_k(x)}{k!} = \sum_{k=0}^{n} \left(\sum_{j=k}^{n} \widehat{p}_{j,k} q_n^{(j)}(0) \right) \phi_k(x).$$

The first relation is the Newton polynomial expansion [113], whereas the second is the Taylor polynomial of $q_n(x)$ in the basis $\phi_k(x)$, $k = 0, \dots, n$.

Proof. It follows from Theorem 3.12. □

Theorem 5.6 (Functional equation). *The functional equation (3.47) for the b. p. s. $(x)_n$ is*

$$\sum_{i=0}^{n-1} i! \widehat{p}_{n,i} \Delta^{n-1} y(x) + n! \widehat{p}_{n,n} y(x) = \sum_{i=0}^{n-1} i! S(n,i) \Delta^{n-1} y(x) + n! S(n,n) y(x) = n! x^n.$$

Finally, from Theorem 3.3, the generating functions are

$$\sum_{k=0}^{\infty} (x)_k \frac{t^k}{k!} = \exp(x \log(t+1)), \quad \sum_{k=0}^{\infty} \phi_k(x) \frac{t^k}{k!} = \exp(x(\exp(t) - 1)). \tag{5.36}$$

The interpolation polynomial related to $f \in C^n[0,1]$, in the case of the lower factorial polynomial, is the well-known Newton polynomial [186].

We note that the sequence of exponential polynomials generates a sequence of linear positive operators, which converges uniformly by Theorem 4.4.

In fact, in [172, Th. 3], it is proved that the sequence

$$L_n[f](x) = \frac{1}{\phi_n(1)} \sum_{k=0}^{n} f\left(\frac{k}{n}\right) \binom{n}{k} \phi_k(x) \phi_{n-k}(1-x)$$

converges uniformly to $f \in C[0,1]$.

From Theorem 4.6, it is possible to accelerate the convergence of $L_n[f]$, that is, the linear operator

$$T_{2n}[f](x) = 2L_{2n}[f](x) - L_n[f](x)$$

has the following property of convergence:

$$T_{2n}[f](x) = f(x) + O\left(\frac{1}{n^2}\right), \quad x \in [0,1], \quad n \to \infty.$$

Some numerical results in Tables 5.1–5.4 confirm this.

5.3 Abel polynomial sequences

Let us consider the numerical sequence $\widehat{b} := (\widehat{b}_i)$:

$$\widehat{b}_0 = 0, \quad \widehat{b}_i = i a^{i-1}, \quad i = 1, 2, \dots, \quad a \in \mathbb{R}, \quad a \neq 0. \tag{5.37}$$

Table 5.1: $f(x) = x^3$ in $x = \frac{1}{2}$.

n	Error L_n	Error T_{2n}
5	0.1615	0.0365
10	0.0990	0.0188
20	0.0589	0.0097
40	0.0343	0.0049

Table 5.2: $f(x) = \sin(x)$ in $x = \frac{1}{2}$.

n	Error L_n	Error T_{2n}
5	0.0254	0.0059
10	0.0156	0.0030
20	0.0093	0.0016
40	0.0054	8.076 E−4

Table 5.3: $f(x) = \cos(x)$ in $x = \frac{1}{2}$.

n	Error L_n	Error T_{2n}
5	0.0465	0.0108
10	0.0286	0.0056
20	0.0171	0.0029
40	0.0100	0.0015

Table 5.4: $f(x) = x^5 + \exp(x)$ in $x = \frac{1}{2}$.

n	Error L_n	Error T_{2n}
5	0.2792	0.0460
10	0.1626	0.0208
20	0.0917	0.0106
40	0.0512	0.0057

From these, we can calculate, by Remark 2.4, the conjugate sequence $b := (b_i)$:

$$b_0 = 0, \quad b_n = (-na)^{n-1}, \quad n = 1, 2, \dots. \tag{5.38}$$

The sequence $\widehat{b}$ generates the power series

$$\bar{f}(t) = \sum_{i=1}^{\infty} \widehat{b}_i \frac{t^i}{i!} = \sum_{i=1}^{\infty} a^{i-1} \frac{t^i}{(i-1)!} = t\exp(at), \tag{5.39}$$

whereas the sequence b generates the formal power series

$$f(t) = \sum_{i=0}^{\infty} b_i \frac{t^i}{i!} = \sum_{i=0}^{\infty} (-ia)^{i-1} \frac{t^i}{i!} = \frac{W(at)}{a}, \tag{5.40}$$

where $W(at)$ is the W-Lambert function [191] defined as

$$W(at)e^{W(at)} = t.$$

Then by Algorithm 2.1.1 we have

$$\begin{cases} \widehat{p}_{n,0} = \delta_{n,0}, \\ \widehat{p}_{n,1} = na^{n-1}, \\ \widehat{p}_{n,k} = \frac{1}{k} \sum_{i=1}^{n-k+1} \binom{n}{i} ia^{i-1} \widehat{p}_{n-i,k-1} = \frac{1}{k!} \sum \binom{n}{v_1,\ldots,v_k} \widehat{p}_{v_1,1} \cdots \widehat{p}_{v_k,1} = \binom{n-1}{k-1} k^{n-k-1} na^{n-k}. \end{cases} \tag{5.41}$$

and

$$\begin{cases} p_{n,0} = \delta_{n,0}, \\ p_{n,1} = (-na)^{n-1}, \\ p_{n,k} = \frac{1}{k} \sum_{i=1}^{n-k+1} \binom{n}{i} (-ia)^{i-1} p_{n-i,k-1} = \frac{1}{k!} \sum \binom{n}{v_1,\ldots,v_k} \widehat{p}_{v_1,1} \cdots \widehat{p}_{v_k,1} = \binom{n-1}{k-1} (-na)^{n-k}. \end{cases} \tag{5.42}$$

From (5.41) and (5.42), we can calculate the associated matrices P and $\widehat{P}$. For example, for $n = 5$, we get

$$\widehat{P} = \begin{pmatrix} 1 & 0 & 0 & 0 & 0 & 0 \\ 0 & 1 & 0 & 0 & 0 & 0 \\ 0 & 2a & 1 & 0 & 0 & 0 \\ 0 & 3a^2 & 6a & 1 & 0 & 0 \\ 0 & 4a^3 & 24a^2 & 12a & 1 & 0 \\ 0 & 5a^4 & 80a^3 & 90a^2 & 20a & 1 \end{pmatrix},$$

$$P = \begin{pmatrix} 1 & 0 & 0 & 0 & 0 & 0 \\ 0 & 1 & 0 & 0 & 0 & 0 \\ 0 & -2a & 1 & 0 & 0 & 0 \\ 0 & 9a^2 & -6a & 1 & 0 & 0 \\ 0 & -64a^3 & 48a^2 & -12a & 1 & 0 \\ 0 & 625a^4 & -500a^3 & 150a^2 & -20a & 1 \end{pmatrix}.$$

With these coefficients, we can define the polynomial sequences

$$p_n(x) = \sum_{k=1}^{n} \binom{n-1}{k-1} (-na)^{n-k} x^k, \tag{5.43}$$

$$\widehat{p}_n(x) = \sum_{k=1}^{n} \binom{n-1}{k-1} k^{n-k-1} na^{n-k} x^k. \tag{5.44}$$

For example, the first six polynomials are

$$\begin{aligned}
\widehat{p}_0(x) &= 1, & p_0(x) &= 1,\\
\widehat{p}_1(x) &= x, & p_1(x) &= x,\\
\widehat{p}_2(x) &= x^2 + 2ax, & p_2(x) &= x^2 - 2ax,\\
\widehat{p}_3(x) &= x^3 + 6ax^2 + 3a^2x, & p_3(x) &= x^3 - 6ax^2 + 9a^2x,\\
\widehat{p}_4(x) &= x^4 + 12ax^3 + 24a^2x^2 + 4a^3x, & p_4(x) &= x^4 - 12ax^3 + 48a^2x^2 - 64a^3x,\\
\widehat{p}_5(x) &= x^5 + 20ax^4 + 90a^2x^3 & p_5(x) &= x^5 - 20ax^4 + 150a^2x^3\\
&\quad + 80a^3x^2 + 5a^4x, & &\quad - 500a^3x^2 + 625a^4x.
\end{aligned}$$

For $a = 1$, their graphics are plotted in Figures 5.3 and 5.4.

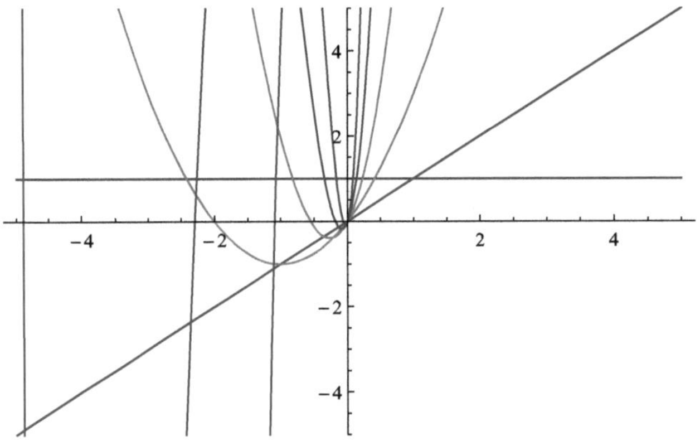

Figure 5.3: Conjugate Abel polynomials.

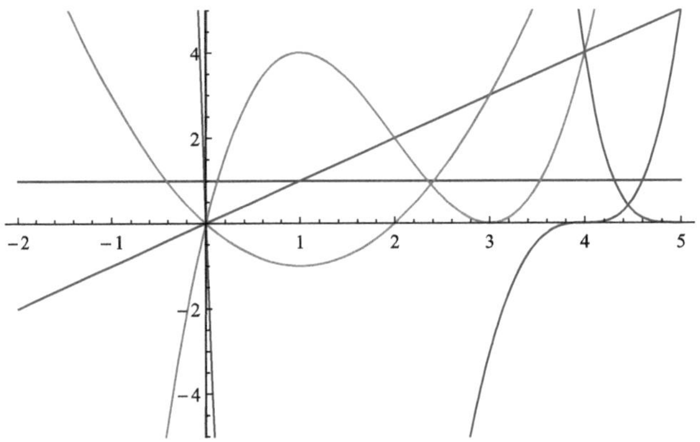

Figure 5.4: Abel polynomials.

Remark 5.6 (Determinantal form). From Theorem 3.7, the determinantal forms of the polynomial sequences (5.43) and (5.44) are

$$p_0(x) = 1,$$
$$p_n(x) = (-1)^{n+1} \begin{vmatrix} x & x^2 & \dots & x^{n-1} & x^n \\ 1 & 2a & \dots & (n-1)a^{n-2} & na^{n-1} \\ 0 & 1 & \dots & \binom{n-2}{1}2^{n-4}(n-1)a^{n-3} & \binom{n-1}{1}2^{n-3}na^{n-2} \\ \vdots & & \ddots & & \vdots \\ 0 & \dots & & 1 & \binom{n-1}{n-2}na \end{vmatrix} \tag{5.45}$$

and

$$\widehat{p}_0(x) = 1,$$
$$\widehat{p}_n(x) = (-1)^{n+1} \begin{vmatrix} x & x^2 & \dots & x^{n-1} & x^n \\ 1 & -2a & 9a^2 & -64a^3 & 625a^4 \\ 0 & 1 & \dots & \binom{n-2}{1}(-(n-1)a)^{n-3} & \binom{n-1}{1}(-na)^{n-2} \\ \vdots & & \ddots & & \vdots \\ 0 & \dots & & 1 & \binom{n-1}{n-2}(-na) \end{vmatrix}. \tag{5.46}$$

Then we have the following:

Theorem 5.7 ([66, Th. 5.12]). *For the b. p. s. $\{p_n\}_n$, we have*

$$p_n(x) = (-1)^{n+1} \begin{vmatrix} x & x^2 & \dots & x^{n-1} & x^n \\ 1 & 2a & \dots & (n-1)a^{n-2} & na^{n-1} \\ 0 & 1 & \dots & \widehat{p}_{n-1,2} & \widehat{p}_{n,2} \\ \vdots & & \ddots & & \vdots \\ 0 & \dots & & 1 & \widehat{p}_{n,n-1} \end{vmatrix} = x(x-na)^{n-1}. \tag{5.47}$$

The sequence $\{p_n\}_n$, by Theorem 5.7, coincides with the polynomial sequence, called the *Abel polynomial sequence* in Roman [166, p. 72].

Theorem 5.8 (Recurrence and differential relations). *For the conjugate b. p. s. $\{x(x-na)^{n-1}\}_n$ and $\{\widehat{p}_n\}_n$, we have:*

$$(x(x-na)^{n-1})' = \sum_{i=1}^{n} \binom{n}{i} (-(i-1)a)^{i-2} x(x-(n-i)a)^{n-i-1}, \tag{5.48}$$

$$\widehat{p}_n'(x) = \sum_{i=1}^{n} \binom{n}{i} (i-1)a^{i-2} \widehat{p}_{n-i}(x), \tag{5.49}$$

$$x(x-na)^{n-1} = x^n - \sum_{k=1}^{n-1} \binom{n-1}{k-1} k^{n-k-1} na^{n-k} x(x-ka)^{k-1}, \tag{5.50}$$

$$\widehat{p}_n(x) = x^n - \sum_{k=1}^{n-1} \binom{n-1}{k-1} (-na)^{n-k} \widehat{p}_k(x), \tag{5.51}$$

$$x(x-(n+1)a)^n = -(n\widehat{c}_1 - x)x(x-na)^{n-1} - \sum_{k=0}^{n-2}\binom{n}{k}\widehat{c}_{n-k}x(x-(k+1)a)^k, \tag{5.52}$$

$$\text{with} \quad \sum_{k=0}^{n}\binom{n}{k}\widehat{c}_k(-(n-k)a)^{n-k-1} = \begin{cases}1, & n=0,\\ 0, & n\geq 1,\end{cases}$$

$$\widehat{p}_{n+1}(x) = -(n\widehat{d}_1 - x)\widehat{p}_n(x) - \sum_{k=0}^{n-2}\binom{n}{k}\widehat{d}_{n-k}\widehat{p}_{k+1}(x), \tag{5.53}$$

$$\text{with} \quad \sum_{k=0}^{n}\binom{n}{k}\widehat{d}_k(n-k)a^{n-k-1} = \begin{cases}1, & n=0,\\ 0, & n\geq 1,\end{cases}$$

$$x(x-(n+1)a)^n = x\sum_{k=0}^{n}\binom{n}{k}(-(n-k)a)^{n-k-1}x(x-ka)^{k-1}, \tag{5.54}$$

$$\widehat{p}_{n+1}(x) = x\sum_{k=0}^{n}\binom{n}{k}(n-k)a^{n-k-1}\widehat{p}_k(x). \tag{5.55}$$

Proof. The proof follows from Theorems 3.1, 3.4, 3.5, 3.6 and previous results. □

Remark 5.7. From the above b. p. s., applying the Proposition 3.2, we get the inverse relations

$$x^n = \sum_{k=1}^{n}\binom{n-1}{k-1}k^{n-k-1}na^{n-k}x(x-ka)^{k-1}, \quad x(x-na)^{n-1} = \sum_{k=1}^{n}\binom{n-1}{k-1}(-na)^{n-k}x^k,$$
$$x^n = \sum_{k=1}^{n}\binom{n-1}{k-1}(-na)^{n-k}\widehat{p}_k(x), \quad \widehat{p}_k(x) = \sum_{k=1}^{n}\binom{n-1}{k-1}k^{n-k-1}na^{n-k}x^k.$$

The binomial identity is

$$(x+y)(x+y-na)^{n-1} = \sum_{k=0}^{n}\binom{n}{k}xy(x-ka)^{k-1}(y-na+ka)^{n-k-1}. \tag{5.56}$$

From the binomial identity, we get the discrete orthogonality for Abel polynomials:

$$\sum_{k=0}^{n}\binom{n}{k}(-1)^{n-k-1}x^2(x-ka)^{k-1}(x+(n-k)a)^{n-k-1} = 0.$$

Remark 5.8 (**δ-operator associated**). The δ-operator associated with the b. p. s. $\{p_n\}$ is

$$Qy = \sum_{i=1}^{\infty} a^{i-1}\frac{y^{(i)}(x)}{(i-1)!} = y'(a+x).$$

The δ-operator associated with the b.p.s $\widehat{p}_n(x)$ is

$$\widehat{Q}y = \sum_{i=1}^{\infty}(-ia)^{i-1}\frac{y^{(i)}(x)}{i!}.$$

Theorem 5.9 (Representation theorem). *For any polynomial $q_n(x)$, the following representations hold:*

$$q_n(x) = \sum_{k=1}^{n} \frac{(Q^k q_n(x))_{x=0}}{k!} x(x-ka)^{k-1}$$
$$= \sum_{k=1}^{n} \left(\sum_{j=k}^{n} \binom{n-1}{j-1} (-na)^{n-j} q_n^{(j)}(ja) \right) x(x-ka)^{k-1}$$

and

$$q_n(x) = \sum_{k=1}^{n} (\widehat{Q}^k q_n(x))_{x=0} \frac{\widehat{p}_k(x)}{k!} = \sum_{k=1}^{n} \left(\sum_{j=k}^{n} \binom{n-1}{j-1} j^{n-j-1} na^{n-j} q_n^{(j)}(0) \right) \frac{\widehat{p}_k(x)}{k!}.$$

Theorem 5.10 (Functional equation). *For the Abel polynomials, the functional equation* (3.47) *becomes*

$$\sum_{i=0}^{n-1} i! \binom{n-1}{n-i} i^{n-i-1} na^{n-i} y^{(n-i)}(ax) + n! y(x) = n! x^n.$$

From Theorem 3.2, the generating functions are

$$\sum_{k=0}^{\infty} x(x-ka)^{k-1} \frac{t^k}{k!} = \exp(xf(t)), \quad \sum_{k=0}^{\infty} \widehat{p}_k(x) \frac{t^k}{k!} = \exp(xt \exp(at)).$$

Finally, we get an interesting interpolation polynomial:

Theorem 5.11. *If $f \in C^n[-1,1]$ and there exists $f^{(n+1)}(x)$ $\forall x \in]-1,1[$, then for $x \in [-1,1]$, the following relation holds:*

$$f(x) = P_n[f](x) + R_n[f](x),$$

where

$$P_n[f](x) = \sum_{k=0}^{n} f^{(k+1)}(ka) \frac{x(x-ka)^{k-1}}{k!},$$
$$R_n[f](x) = \int_{-1}^{1} K_n(x,t) f^{(n+1)}(t) dt$$

with

$$K_n(x,t) = \frac{1}{n!} \left[(x-t)_+^n - \sum_{k=0}^{n} D^{k+1} [(x-t)_+^n]_{x=0} \right].$$

Moreover, we have

$$P_n^{(k+1)}[f](ka) = f^{(k+1)}(ka), \quad k = 0, \dots, n.$$

A more general Abel–Gontscharoff-type interpolation is considered in [9, p. 172] and references therein.

5.4 Binomial Laguerre polynomials

Let us consider the numerical sequence $\widehat{b} = (\widehat{b}_i)$:

$$\widehat{b}_0 = 0, \quad \widehat{b}_i = -i!, \quad i = 1, 2, \dots. \tag{5.57}$$

From the Algorithm 2.1.2, we get the conjugate sequence $b = (b_i)$:

$$b_0 = 0, \quad b_i = -i!, \quad i = 1, 2, \dots. \tag{5.58}$$

Since the sequences b and $\widehat{b}$ coincide, for the power series $f(t)$ and $\overline{f}(t)$, we have that

$$f(t) = \overline{f}(t) = \sum_{k=0}^{\infty} (-k!) \frac{t^k}{k!} = \frac{t}{t-1}. \tag{5.59}$$

From (2.1), we get

$$\begin{cases} p_{n,0} = \widehat{p}_{n,0} = \delta_{n,0}, \\ p_{n,1} = \widehat{p}_{n,1} = -n!, \\ p_{n,k} = \widehat{p}_{n,k} = \frac{1}{k} \sum_{i=1}^{n-k+1} \binom{n}{i} (-i!) \widehat{p}_{n-i,k-1} = (-1)^k \frac{n!}{k!} \binom{n-1}{n-k}, \end{cases} \tag{5.60}$$

where the last identity follows from [164, p. 195].

The coefficients $p_{n,k} = \widehat{p}_{n,k}$ are the elements of the conjugate matrices P and $\widehat{P}$. These coefficients are integers, as can be proved by mathematical induction from (5.60).

For example, for $n = 5$, we have:

$$P = \widehat{P} = \begin{pmatrix} 1 & 0 & 0 & 0 & 0 & 0 \\ 0 & -1 & 0 & 0 & 0 & 0 \\ 0 & -2 & 1 & 0 & 0 & 0 \\ 0 & -6 & 6 & -1 & 0 & 0 \\ 0 & -24 & 36 & -12 & 1 & 0 \\ 0 & -120 & 240 & -120 & 20 & -1 \end{pmatrix}.$$

From (3.2), we get the polynomial sequences $\{p_n\}_n = \{\widehat{p}_n\}_n$:

$$\widehat{p}_n(x) = p_n(x) = \sum_{k=0}^{n} (-1)^k \frac{n!}{k!} \binom{n-1}{n-k} x^k. \tag{5.61}$$

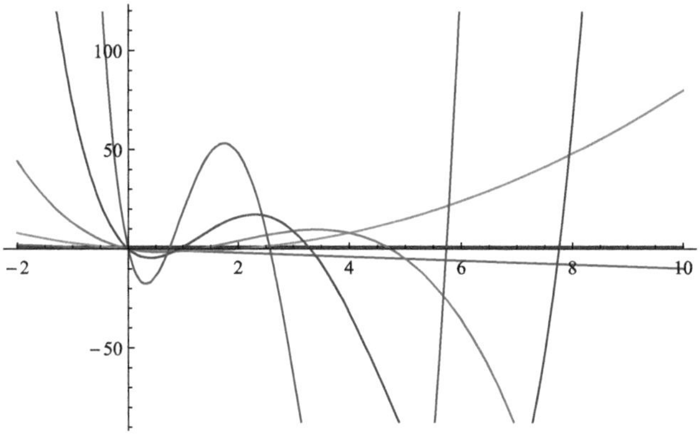

Figure 5.5: Binomial Laguerre polynomials.

For example, the first six polynomials are

$$\begin{aligned}
p_0(x) &= 1,\\
p_1(x) &= -x,\\
p_2(x) &= x^2 - 2x,\\
p_3(x) &= -x^3 + 6x^2 - 6x,\\
p_4(x) &= x^4 - 12x^3 + 36x^2 - 24x,\\
p_5(x) &= -x^5 + 20x^4 - 120x^3 + 240x^2 - 120x.
\end{aligned}$$

We display the graphics in Figure 5.5.

The coefficient of x^n is $(-1)^n$, whereas the coefficient of x is $-n!$

Remark 5.9. We note that these polynomials are related to the generalized Laguerre polynomials [163, p. 200]. In fact,

$$p_n(x) = n!L_n^{(-1)}(x), \tag{5.62}$$

where $L_n^{(\alpha)}(x)$ is the generalized Laguerre polynomial, as defined in [191, p. 100] and [163, p. 201]:

$$L_n^{(\alpha)}(x) = \sum_{k=0}^{n} \frac{(-1)^k (1+\alpha)_n x^k}{k!(n-k)!(1+\alpha)_k}, \tag{5.63}$$

where $\forall \alpha \neq 0$, $(\alpha)_0 = 1$, $(\alpha)_n = \alpha(1+\alpha)(2+\alpha)\cdots(n-1+\alpha)$, $\forall n \geq 1$.

The case $\alpha = -1$ of generalized Laguerre polynomials is not much studied in the literature [167, p. 115] and above-cited references. It is interesting to note that this case, that is, $\{p_n\}_n$ with $p_n(x) = n!L_n^{(-1)}(x)$, is the unique b. p. s. in the class of generalized Laguerre polynomials.

Now we get some unknown properties of binomial Laguerre polynomials

Theorem 5.12 (Recurrence relations). *For the b. p. s.* $\{p_n\} = \{\widehat{p}_n\}$, *given that* $P = \widehat{P}$, *we have*

$$p_n(x) = (-1)^n \left[x^n - \sum_{k=0}^{n-1} p_{n,k} p_k(x) \right], \tag{5.64}$$

$$p_{n+1}(x) = (2n - x)p_n(x) + \sum_{k=0}^{n-2} \binom{n}{k} \widehat{c}_{n-k} p_{k+1}(x) \tag{5.65}$$

$$\text{with} \quad \begin{cases} \widehat{c}_0 = 1, \\ \widehat{c}_n = -\sum_{k=0}^{n-1} \binom{n}{k} \widehat{c}_k (n-k+1)!, \end{cases}$$

$$p_{n+1}(x) = -x \sum_{k=0}^{n} \binom{n}{k} (n-k+1)! p_k(x). \tag{5.66}$$

Theorem 5.13 (Determinantal forms). *For the b. p. s. associated with the sequence* $\widehat{b}$, *we have the following determinantal forms:*

$$\widehat{p}_0(x) = p_0(x) = 1,$$

$$\widehat{p}_n(x) = p_n(x) = - \begin{vmatrix} x & x^2 & \cdots & x^{n-1} & x^n \\ -1 & -2 & \cdots & -(n-1)! & -n! \\ 0 & 1 & \cdots & \widehat{p}_{n-1,2} & \widehat{p}_{n,2} \\ \vdots & & \ddots & & \vdots \\ 0 & \cdots & & (-1)^n & \widehat{p}_{n,n-1} \end{vmatrix}, \tag{5.67}$$

and

$$\widehat{p}_0(x) = p_0(x) = 1,$$

$$\widehat{p}_n(x) = p_n(x) = - \begin{vmatrix} x & -1 & 0 & \cdots & 0 & 0 \\ 0 & x-2 & -1 & \cdots & 0 & 0 \\ 0 & -\binom{2}{0} c_2 & x-4 & \cdots & 0 & 0 \\ \vdots & & & \ddots & & \\ 0 & -\binom{n-1}{0} c_{n-1} & -\binom{n-1}{1} c_{n-2} & \cdots & x-2(n-1) & -1 \\ 0 & -\binom{n}{0} c_n & -\binom{n}{1} c_{n-1} & & \cdots & x-2n \end{vmatrix}. \tag{5.68}$$

Theorem 5.14 (Differential relation). *For the b. p. s.* $\{p_n\}$, *the following differential relation holds:*

$$p_n'(x) = -\sum_{i=1}^{n} \binom{n}{i} i! p_{n-i}(x). \tag{5.69}$$

Proof. The proof follows from Theorems 3.1 and previous results. □

Theorem 5.15 (Properties of zeros). *The zeros of $p_n(x)$ satisfy:*
1. *All zeros are real, simple, and nonnegative;*
2. *The rational zeros are $x = 0$, and for $n > 2$, there are not other rational zeros.*

Proof. From (1) and from [163, p. 203], we get

$$p_n'(x) = -\frac{1}{n!}L_{n-1}^{(0)}(x) = -\frac{1}{n!}L_{n-1}(x).$$

Then the result follows from known properties of $L_{n-1}(x)$ and by differential calculus.

Condition (2) follows from the known properties of polynomials with integer coefficients [52, p. 50]. □

Theorem 5.16 (Semiorthogonality). *The Laguerre b. p. s. satisfies*

$$\int_0^{+\infty} \frac{\exp(-x)}{x} p_n(x)p_m(x)dx = 0 \quad \text{for } m \neq n \neq 0. \tag{5.70}$$

Proof. For $m \neq n \neq 0$, we have

$$(m-n)\frac{\exp(-x)}{x}p_n(x)p_m(x) = D\{\exp(-x)[p_m(x)p_n'(x) - p_n(x)p_m'(x)]\},$$

from which it follows

$$(m-n)\int_a^b \frac{\exp(-x)}{x}p_n(x)p_m(x)dx = \exp(-x)[p_m(x)p_n'(x) - p_n(x)p_m'(x)].$$

Now observing that the product of $\exp(-x)$ and any polynomial tends to zero when $x \to \infty$ and that $p_m(0) = p_n(0) = 0$ for $m \neq n \neq 0$, the result follows. □

Remark 5.10. The Laguerre b. p. s. is not an orthogonal sequence. Consequently, we say that $\{p_n\}_n$ is semiorthogonal.

Corollary 5.1. *The generalized Laguerre polynomials are orthogonal if* $\mathrm{Re}(\alpha) > -1$.

Remark 5.11. From Proposition 3.2, we get the inverse relations

$$x^n = \sum_{k=0}^{n}(-1)^k\frac{n!}{k!}\binom{n-1}{n-k}p_k(x), \quad p_n(x) = \sum_{k=0}^{n}(-1)^k\frac{n!}{k!}\binom{n-1}{n-k}x^k. \tag{5.71}$$

Theorem 5.17 (Binomial identity and discrete orthogonality). *The binomial identity is*

$$p_n(x+y) = \sum_{k=0}^{n}\binom{n}{k}p_k(x)p_{n-k}(x), \tag{5.72}$$

and we get the discrete orthogonality

$$\sum_{k=0}^{n}\binom{n}{k}p_k(x)p_{n-k}(-x) = 0.$$

Now we consider the linear functional

$$\hat{L}(p) = -\int_0^{+\infty} \exp(-t)p'(t)dt \quad \forall p \in \mathcal{P}. \tag{5.73}$$

The result is

$$\hat{L}(x^i) = -\int_0^{+\infty} \exp(-t)it^{i-1}dt = -i! = \hat{b}_i.$$

Then we have

$$\hat{L}^2(x^i) = 2p_{i,2}, \quad \hat{L}^k(x^i) = k!p_{i,k},$$

and therefore, from Theorem 3.9,

$$\hat{L}^k(p_n) = n!\delta_{n,k}, \quad k = 0, 1, \dots.$$

Theorem 5.18 (δ-operator associated). *The δ-operator associated with the b. p. s.* $\{p_n\}_n$ *is*

$$Qy = -\int_0^{\infty} e^{-t}y'(x+t)dt. \tag{5.74}$$

Proof. Observing that

$$y'(x+t) = \sum_{i=0}^{\infty} \frac{y^{(i+1)}(x)}{i!}t^i,$$

relation (5.74) follows from

$$Qy = -\sum_{i=1}^{\infty} y^{(i)}(x)$$

after easy calculations. □

Theorem 5.19 (Representation theorem). *For any* $q_n(x) \in \mathcal{P}_n$*, the following representation holds:*

$$q_n(x) = \sum_{k=1}^{n} \frac{(Q^k q_n(x))_{x=0}}{k!} L_k^{(-1)}(x) = \sum_{k=1}^{n} \left(\sum_{j=k}^{n} (-1)^k \binom{n-1}{j-k} q_n^{(j)}(0) \right) \frac{L_k^{(-1)}(x)}{k!}. \tag{5.75}$$

Theorem 5.20 (Generating function). *The generating function of the binomial Laguerre polynomials is*

$$\sum_{k=0}^{\infty} p_k(x)\frac{t^k}{k!} = \exp\left(x\frac{t}{t-1}\right). \tag{5.76}$$

Now we can consider the polynomial sequence $\{\varphi_n\}_n$ defined by

$$\varphi_n(x) = n!L_n^{(-1)}(-x) = p_n(-x), \quad n = 0, 1, \ldots; \tag{5.77}$$

therefore, from (5.61), we get

$$\varphi_n(x) = \sum_{k=1}^{n} \frac{n!}{k!} \binom{n-1}{n-k} x^k, \quad n \geq 1.$$

It is known in the literature [172, Ex. 4.2], [164, p. 94], [42, p. 165] that for $h(t) = \frac{t}{1-t}$, the result is

$$\exp(xh(t)) = \sum_{n=0}^{\infty} \varphi_n(x) \frac{t^n}{n!}. \tag{5.78}$$

We have that $\{\varphi_n\}_n$ is the b. p. s. with $b_0 = 0$ and $b_i = i!$ for $i = 1, \ldots$.

Moreover, $\varphi_n(x) \geq 0$ for $x \geq 0$, and therefore we can consider the linear operator

$$L_n[f](x) = \frac{1}{\varphi_n(1)} \sum_{k=0}^{n} \binom{n}{k} \varphi_n(x) \varphi_{n-k}(1-x) f\left(\frac{k}{n}\right). \tag{5.79}$$

It is proved [172, Ex. 4.2] that this operator satisfies the hypotheses of Theorems 4.2, 4.3, 4.4, and hence

$$\lim_{n\to\infty} L_n[f](x) = f(x) \quad \text{uniformly for} \quad x \in (0, 1).$$

5.5 Central factorial polynomials

Let us consider the sequence $\widehat{b} := (\widehat{b}_i)_i$ with

$$\widehat{b}_0 = 0 \quad \widehat{b}_n = \begin{cases} 0 & \text{if } n \text{ is even,} \\ \frac{1}{2^{n-1}} & \text{if } n \text{ is odd.} \end{cases} \tag{5.80}$$

Then we can calculate its conjugate sequence $b := (b_i)_i$:

$$b_0 = 0, \quad b_n = \begin{cases} 0 & \text{if } n \text{ is even,} \\ -\sum_{k=0}^{n-1} b_k \widehat{b}_{n,k} & \text{if } n \text{ is odd.} \end{cases} \tag{5.81}$$

Hence, we can consider the power series

$$\overline{f}(t) = \sum_{k=0}^{\infty} \widehat{b}_k \frac{t^k}{k!} = \sum_{i=0}^{\infty} \frac{1}{2^{2i}} \frac{t^{2i+1}}{(2i+1)!} = \exp\left(\frac{t}{2}\right) - \exp\left(-\frac{t}{2}\right), \quad f(t) = \sum_{k=0}^{\infty} b_k \frac{t^k}{k!}. \tag{5.82}$$

As a result, from Algorithm 2.1.1, we get

$$\begin{cases} \widehat{p}_{n,0} = \delta_{n,0}, & \\ \widehat{p}_{n,1} = \frac{1}{2^{n-1}} & \text{if } n \text{ is odd,} \\ \widehat{p}_{n,1} = 0 & \text{if } n \text{ is even,} \\ \widehat{p}_{n,k} = \frac{1}{k}\sum_{i=0}^{n-k+1}\binom{n}{i}\frac{1}{2^{i-1}}\widehat{p}_{n-i,k-1} & \text{if } n \text{ and } k \text{ have the same parity,} \quad n \geq k, \\ 0 & \text{otherwise,} \end{cases} \tag{5.83}$$

and

$$\begin{cases} p_{n,0} = \delta_{n,0}, & \\ p_{n,1} = b_n & \text{if } n \text{ is odd,} \\ p_{n,1} = 0 & \text{if } n \text{ is even,} \\ p_{n,k} = \frac{1}{k}\sum_{i=0}^{n-k+1}\binom{n}{i}b_i p_{n-i,k-1} & \text{if } n \text{ and } k \text{ have the same parity,} \quad n \geq k, \\ 0 & \text{otherwise.} \end{cases}$$

Hence, we have the conjugate matrices $P := (p_{i,k})$ and $\widehat{P} := (\widehat{p}_{i,k})$.

For example, for $n = 5$, we get

$$\widehat{P} = \begin{pmatrix} 1 & 0 & 0 & 0 & 0 & 0 \\ 0 & 1 & 0 & 0 & 0 & 0 \\ 0 & 0 & 1 & 0 & 0 & 0 \\ 0 & \frac{1}{4} & 0 & 1 & 0 & 0 \\ 0 & 0 & 1 & 0 & 1 & 0 \\ 0 & \frac{1}{16} & 0 & \frac{5}{2} & 0 & 1 \end{pmatrix}, \quad P = \begin{pmatrix} 1 & 0 & 0 & 0 & 0 & 0 \\ 0 & 1 & 0 & 0 & 0 & 0 \\ 0 & 0 & 1 & 0 & 0 & 0 \\ 0 & -\frac{1}{4} & 0 & 1 & 0 & 0 \\ 0 & 0 & -1 & 0 & 1 & 0 \\ 0 & \frac{9}{16} & 0 & -\frac{5}{2} & 0 & 1 \end{pmatrix}.$$

The entries of the conjugate matrices P, $\widehat{P}$ present some interesting regularities:

Theorem 5.21. *For the above conjugate binomial-type matrices, the following conditions hold:*

$$\widehat{p}_{k+1,k} = p_{k+1,k} = 0, \quad k = 0, 1, \ldots, \tag{5.84}$$

$$\widehat{p}_{2k,k-1} = p_{2k,k-1} = 0, \quad k = 1, 2, \ldots. \tag{5.85}$$

Proof. We proceed by induction on k.

For relation (5.84), if $k = 0$, the result follows from (5.83).

Then by induction

$$\begin{aligned} p_{k+1,k} &= \frac{1}{k}\sum_{k=1}^{2}\binom{k+1}{i} b_i p_{k+1-i,k-1} \\ &= \frac{1}{k}\left[\binom{k+1}{1} b_1 p_{k,k-1} + \binom{k+1}{2} b_2 p_{k-1,k-1}\right] = 0, \end{aligned}$$

given that $b_0 = b_2 = 0$ from (5.83) and $p_{k,k-1} = 0$ by induction hypothesis.

By analogy, we get (5.85). □

Now from the conjugate matrices, we can define the b. p. s.:

$$p_n(x) = \sum_{k=0}^{n} p_{n,k} x^k, \quad \widehat{p}_n(x) = \sum_{k=0}^{n} \widehat{p}_{n,k} x^k.$$

For example, the first six polynomials are

$$\begin{aligned}
\widehat{p}_0 &= 1, & p_0 &= 1,\\
\widehat{p}_1(x) &= x, & p_1(x) &= x,\\
\widehat{p}_2(x) &= x^2, & p_2(x) &= x^2,\\
\widehat{p}_3(x) &= x^3 + \frac{1}{4}x, & p_3(x) &= x^3 - \frac{1}{4}x,\\
\widehat{p}_4(x) &= x^4 + x^2, & p_4(x) &= x^4 - x^2,\\
\widehat{p}_5(x) &= x^5 + \frac{5}{2}x^3 + \frac{1}{16}x, & p_5(x) &= x^5 - \frac{5}{2}x^3 + \frac{9}{16}x.
\end{aligned}$$

The graphics of these polynomials are shown in Figures 5.6 and 5.7.

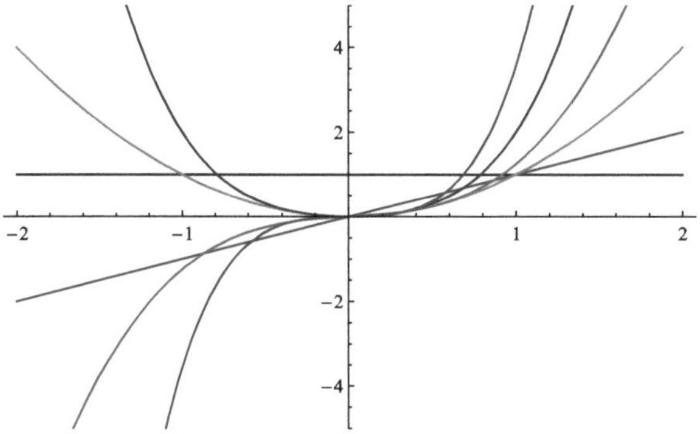

Figure 5.6: Carlitz–Riordan polynomials.

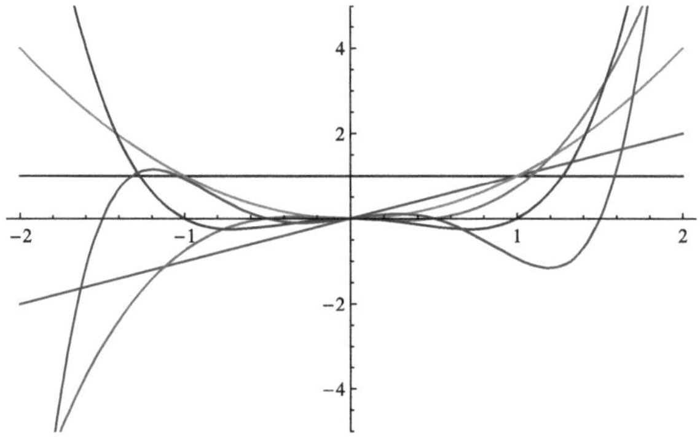

Figure 5.7: Central factorial polynomials.

In order to give a closed form to the b. p. s. $\{p_n\}$ and $\{\widehat{p}_n\}$, we consider the δ-operator associated with $\{p_n\}$. From Theorem 3.11, we get

$$Qy = \sum_{i=1}^{\infty} \widehat{b}_i \frac{y^{(i)}}{i!} = \sum_{i=0}^{\infty} \frac{1}{2^{2i}} \frac{y^{(2i+1)}}{(2i+1)!} = y\left(x+\frac{1}{2}\right) - y\left(x-\frac{1}{2}\right). \tag{5.86}$$

Recalling the central difference operator [164, p. 212], [186, p. 7], from (5.86), we have

$$Qy = \delta y. \tag{5.87}$$

So the b. p. s. $\{p_n\}$ satisfies

$$\begin{cases} \delta p_n(x) = np_{n-1}(x), & n = 1, 2, \ldots, \\ p_0(x) = 1, \end{cases} \tag{5.88}$$

and by induction we have

$$\begin{aligned} p_1(x) &= x, \\ p_2(x) &= x^2, \\ p_3(x) &= x\left(x+\frac{1}{2}\right)\left(x-\frac{1}{2}\right), \\ &\cdots\cdots\cdots \\ p_n(x) &= x\left(x+\frac{n}{2}-1\right)\left(x+\frac{n}{2}-2\right)\cdots\left(x-\frac{n}{2}+1\right). \end{aligned}$$

These polynomials are denoted in the literature [164, p. 212], [186, p. 8] by $x^{[n]}$, that is,

$$p_n(x) = x\left(x+\frac{n}{2}-1\right)\left(x+\frac{n}{2}-2\right)\cdots\left(x+\frac{n}{2}-n+1\right) \equiv x^{[n]}, \tag{5.89}$$

and are called *central factorial polynomials.*

The coefficients of $p_n(x)$ in the canonical bases $\{x^k\}$ are denoted in the literature [164, p. 212] by $t(n,k)$, that is,

$$x^{[n]} \equiv p_n(x) = \sum_{k=0}^{n} p_{n,k} x^k = \sum_{k=0}^{n} t(n,k) x^k. \tag{5.90}$$

Given the P and $\widehat{P}$ inverse matrices, from (5.90), we get the inverse relation

$$x^n = \sum_{k=0}^{n} \widehat{p}_{n,k} x^{[k]}. \tag{5.91}$$

In this contest, the coefficients $\widehat{p}_{n,k}$ are denoted [164, p. 212] by $T(n,k)$, that is,

$$x^n = \sum_{k=0}^{n} T(n,k)x^{[k]}. \tag{5.92}$$

From (5.91) and (5.92), the numbers $t(n,k)$ and $T(n,k)$ are analogous to the Stirling numbers. Additional properties of these numbers are in [164, p. 212].

The δ-operator allows us to prove a representation theorem for polynomials:

Theorem 5.22 (Representation theorem). *For any polynomial $q_n(x)$, the representation theorem is*

$$q_n(x) = \sum_{k=0}^{n} \sum_{j=0}^{k} \binom{k}{j} (-1)^j q_n\left(-j + \frac{1}{2}k\right)x^{[k]}. \tag{5.93}$$

Proof. From Theorem 3.12, according to [164, p. 212], we have

$$\delta^k q_n(x) = \sum_{j=0}^{k} \binom{k}{j} (-1)^j q_n\left(x - j + \frac{1}{2}k\right).$$

Hence, the result follows. □

Relation (5.93) is known as the Newton formula for polynomials [164, p. 213].

Finally, the central factorial polynomials have a role in the interpolation problem. In fact, for a regular real function f, the interpolation polynomial is

$$P_n[f](x) = \sum_{k=0}^{n} \delta^k f(x)\frac{x^{[k]}}{k!} \tag{5.94}$$

with the remainder

$$R_n[f](x) = \int_0^1 K_n(x,t)f^{(n+1)}(t)dt,$$

where

$$K_n(x,t) = \frac{1}{n!}\left[(x-t)_+^n - \sum_{k=0}^{n} \delta^k((x-t)_+^n)_{x=0}\right].$$

This interpolatory formula is well known; for example, we can find it in [186, p. 33].

The b. p. s. $\{\widehat{p}_k\}_k$ is nonnegative in $[0,+\infty[$. For this reason, it can be used in operators approximation theory both in finite and in semiinfinite intervals.

As regards the determinantal form for central factorial polynomials, we get the following:

Theorem 5.23 (Determinantal form). *From* (3.31), *the b. p. s. associated with the sequence* $(\widehat{b}_i)_i$ *is*

$$\widehat{p}_0(x) = 1,$$
$$\widehat{p}_n(x) = -\begin{vmatrix} x & x^2 & \dots & x^{n-1} & x^n \\ 1 & 0 & \dots & \frac{1}{2^{n-2}} & 0 \\ 0 & \widehat{p}_{2,2} & \dots & 0 & \widehat{p}_{n,2} \\ \vdots & \ddots & & \vdots & \vdots \\ 0 & \dots & & 1 & 0 \end{vmatrix} \quad \text{if } n \text{ is even,}$$
$$\widehat{p}_n(x) = \begin{vmatrix} x & x^2 & \dots & x^{n-1} & x^n \\ 1 & 0 & \dots & 0 & \frac{1}{2^{n-1}} \\ 0 & \widehat{p}_{2,2} & \dots & \widehat{p}_{n-1,2} & 0 \\ \vdots & \ddots & & \vdots & \vdots \\ 0 & \dots & & 1 & 0 \end{vmatrix} \quad \text{if } n \text{ is odd.} \tag{5.95}$$

The b. p. s. associated with $(b_i)_i$ *is*

$$p_0(x) = 1,$$
$$p_n(x) = -\begin{vmatrix} x & x^2 & \dots & x^{n-1} & x^n \\ 1 & 0 & \dots & p_{n-1,1} & 0 \\ 0 & 1 & \dots & 0 & p_{n,2} \\ \vdots & \ddots & & \vdots & \vdots \\ 0 & \dots & & 1 & 0 \end{vmatrix} \quad \text{if } n \text{ is even,}$$
$$p_n(x) = \begin{vmatrix} x & x^2 & \dots & x^{n-1} & x^n \\ 1 & 0 & \dots & 0 & p_{n,1} \\ 0 & 1 & \dots & \widehat{p}_{n-1,2} & 0 \\ \vdots & \ddots & & \vdots & \vdots \\ 0 & \dots & & 1 & 0 \end{vmatrix} \quad \text{if } n \text{ is odd.} \tag{5.96}$$

Moreover, we get some recurrence and differential relations:

Theorem 5.24 (Recurrence and differential relations). *For the conjugate b. p. s.* $\{x^{[n]}\}_n$, $\{\widehat{p}_n\}_n$, *we have*

$$(x^{[n]})' = \sum_{i=1}^{n} \binom{n}{i} b_{i-1} x^{[n-i]}, \tag{5.97}$$

$$\widehat{p}_n'(x) = \sum_{i=1}^{n} \binom{n}{i} \widehat{b}_{i-1} \widehat{p}_{n-i}(x), \tag{5.98}$$

$$x^{[n]} = x^n - \sum_{k=0}^{n-1} \widehat{p}_{n,k} x^{[k]}, \tag{5.99}$$

$$\widehat{p}_n(x) = x^n - \sum_{k=0}^{n-1} p_{n,k}\widehat{p}_k(x), \tag{5.100}$$

$$x^{[n+1]} = -(n\widehat{c}_1 - x)x^{[n]} - \sum_{k=0}^{n-2} \binom{n}{k} \widehat{c}_{n-k} x^{[k+1]}, \tag{5.101}$$

$$\text{with} \quad \sum_{k=0}^{n} \binom{n}{k} \widehat{c}_k b_{n-k+1} = \delta_{n,0},$$

$$\widehat{p}_{n+1}(x) = -(n\widehat{d}_1 - x)\widehat{p}_n(x) - \sum_{k=0}^{n-2} \binom{n}{k} \widehat{d}_{n-k}\widehat{p}_{k+1}(x), \tag{5.102}$$

$$\text{with} \quad \sum_{k=0}^{n} \binom{n}{k} \widehat{d}_k \widehat{b}_{n-k+1} = \delta_{n,0},$$

$$x^{[n+1]} = x \sum_{k=0}^{n} \binom{n}{k} p_{n-k} x^{[k]}, \tag{5.103}$$

$$\widehat{p}_{n+1}(x) = x \sum_{k=0}^{n} \binom{n}{k} \widehat{p}_{n-k}\widehat{p}_k(x). \tag{5.104}$$

Proof. The proof follows by Theorems 3.1, 3.4, 3.5, 3.6 and the previous results. □

Part III: **Appell polynomial sequences**

Je m'occupe, dans ce Memoir, de certains polynômes en x formant une suite

$$A_0, A_1, \ldots, A_n, \ldots$$

dont le terme A_n est un polynôme de degre n et dans laquelle deux termes consècutifes sont liès par la relation

$$\frac{d\,A_n}{dx} = nA_{n-1}.$$

M. P. Appell (1880)

6 Appell polynomial sequences

Abstract: With the techniques introduced in the previous chapters, we consider the set of Appell sequences. General properties, recurrence relations, determinantal forms, localization of zeros, and differential equations are considered, too. Finally, links with linear functionals and Rota's theory are discussed.

6.1 Introduction

In 1880, Appell [17] introduced a widely studied sequence of nth-degree polynomials $\{A_n\}_{n\in\mathbb{N}}$ satisfying the differential relation

$$A_n'(x) = nA_{n-1}(x), \quad n = 1, 2, \ldots .$$

These polynomial sequences are called Appell polynomials and have many applications not only in different branches of Mathematics [18, 67, 76, 177, 192], but also in theoretical Physics and Chemistry [126, 127]. In 1936, an initial bibliography was provided by Davis [75]. In 1939, Sheffer [179] extended the class of Appell polynomials and called these polynomials A-type zero, but currently we call them Sheffer polynomials. Sheffer also noticed the similarities between Appell polynomials and the umbral calculus, introduced in the second half of the 19th century with the works of Mathematicians, such as Sylvester, Cayley, and Blissard (for examples see [23]). The Sheffer theory is mainly based on a formal power series. In 1941, Steffensen [185] published a theory on Sheffer polynomials also based on formal power series. Afterwards Mullin, Roman, and Rota [142, 166, 167], gave a beautiful theory of umbral calculus, including Sheffer polynomials, using operators method. Recently, Di Bucchianico and Loeb [79] collected a series of old and new works related to Appell polynomial sequences. In the last few years, the main interest is in finding new representations of Appell polynomials by means of matrix theory. For instance, Lehmer [125] illustrated six different approaches for the representation of the sequence of Bernoulli polynomials, which are a special case of Appell sequences. After [43, 44], there is a new characterization of Bernoulli polynomials, as determinantal form, and this idea is extended to more general Appell sequences in [60, 62, 214]. Now we will give an elementary theory based on matrix calculus. Finally, generating function and links with the linear functional are sketched.

https://doi.org/10.1515/9783110652925-006

6.2 Definition and characterizations

Let A be an Appell-type matrix (2.9), that is, there exists a sequence $(a_i)_i$ of elements of $\mathbb{K}$ such that $a_0 \neq 0$, and if

$$\begin{cases} a_{i,0} = a_i, & i = 0, 1, \dots, \\ a_{i,j} = \binom{i}{j} a_{i-j}, & j = 1, 2, \dots, \quad i \geq j, \\ a_{i,j} = 0, & i < j, \end{cases} \tag{6.1}$$

we have the matrix $A := (a_{i,j})_{i,j=0,1,\dots}$.

This matrix allows us to give the following definition.

Definition 6.1. The polynomial sequence $\{a_n\}_{n\in\mathbb{N}}$ defined by

$$\begin{cases} a_0(x) = a_{0,0}, \\ a_1(x) = a_{1,0} + a_{1,1}x, \\ \vdots \\ a_n(x) = a_{n,0} + a_{n,1}x + \cdots + a_{n,n}x^n, \\ \vdots \end{cases} \tag{6.2}$$

with $a_{i,j}$, defined in (6.1), is called an *Appell polynomial sequence (A. p. s.)* for the matrix A or for the sequence $(a_i)_i$ (in the following, A. p. s.).

Remark 6.1. We note explicitly that by (6.1), $a_n(x)$ can be written as

$$a_n(x) = a_n + \binom{n}{1} a_{n-1}x + \cdots + a_0 x^n = \sum_{i=0}^{n} \binom{n}{i} a_{n-i}x^i, \tag{6.3}$$

where the coefficients a_i, $i = 0, \dots, n$, are a priori assigned and hence independent from the degree n.

Moreover, we can rewrite it as

$$a_n(x) = \sum_{j=0}^{n} \binom{n}{j} a_{n-j}(0)x^j. \tag{6.4}$$

Proposition 6.1. *An A. p. s. $\{a_n\}$ is nonnegative in $[0, +\infty[$, that is, $a_n(x) \geq 0$ $\forall x \in [0, +\infty[$ if the numerical sequence (a_n) is nonnegative, that is,*

$$a_n \geq 0 \quad \forall n \in \mathbb{N}. \tag{6.5}$$

Proof. If (6.5) holds, then the result follows from (6.3).

Vice versa, if $a_n(x) \geq 0$ $\forall x \in [0, +\infty[$, from (6.4), it follows that

$$0 \leq a_n(0) = a_n \quad \forall n \in \mathbb{N}. \qquad \square$$

Let $\widehat{A} = (\widehat{a}_{i,k})$ be the conjugate matrix of A as in Algorithm 2.2.2. Given an Appell-type matrix $\widehat{A}$, we can consider the A. p. s. $\{\widehat{a}_n\}_n$ defined by

$$\begin{cases} \widehat{a}_0(x) = \widehat{a}_{0,0}, \\ \widehat{a}_1(x) = \widehat{a}_{1,0} + \widehat{a}_{1,1}x, \\ \quad\vdots \\ \widehat{a}_n(x) = \widehat{a}_{n,0} + \widehat{a}_{n,1}x + \cdots + \widehat{a}_{n,n}x^n. \\ \quad\vdots \end{cases} \tag{6.6}$$

The sequences $\{\widehat{a}_n\}_n$ and $\{a_n\}_n$ are called *conjugate A. p. s.*

We recall that from (2.11), we have

$$\widehat{a}_{n,k} = \binom{n}{k}\widehat{a}_{n-k}$$

with

$$\forall n \in \mathbb{N} \quad : \quad \sum_{k=0}^{n}\binom{n}{k}a_k\widehat{a}_{n-k} = \delta_{n,0}, \tag{6.7}$$

where $\delta_{i,j}$ is the Kronecker symbol.

Moreover, the matrices A and $\widehat{A}$ are inverse of each other.

We note that

$$\sum_{k=0}^{n}\binom{n}{k}a_k\widehat{a}_{n-k} = \sum_{k=0}^{n}\binom{n}{k}a_{n-k}\widehat{a}_k$$

is called the *product of convolution* or *binomial convolution* of the sequences (a_k) and $(\widehat{a}_k)$.

Then for the conjugate A. p. s. $\{a_n\}_n$ and $\{\widehat{a}_n\}_n$, we get the following important results.

Proposition 6.2. *For the conjugate A. p. s. $\{a_n\}_n$ and $\{\widehat{a}_n\}_n$, we have*

$$a_n(\widehat{a}_n(x)) = \widehat{a}_n(a_n(x)) = x^n, \quad n = 1, 2, \ldots, \tag{6.8}$$

where

$$a_n(\widehat{a}_n(x)) = \sum_{k=0}^{n}\binom{n}{k}a_{n-k}\widehat{a}_k(x) = \sum_{k=0}^{n}\binom{n}{k}\widehat{a}_{n-k}a_k(x) = \widehat{a}_n(a_n(x))$$

is the umbral composition [166, *p. 41*], [167, *p. 120*].

Proof. Given that the matrices A and $\widehat{A}$ are inverse of each other, the result easily follows. □

Theorem 6.1. *([60]) A polynomial sequence $\{a_n\}_n$ is an A. p. s. if and only if*

$$
\begin{aligned}
&a_0(x) = a_0 \neq 0, \\
&a_n'(x) = \frac{d}{dx} a_n(x) \equiv Da_n(x) = n a_{n-1}(x), \quad n = 1, \dots, \\
&a_n(0) = a_n, \quad n = 1, 2, \dots .
\end{aligned}
\tag{6.9}
$$

Proof. By integration of (6.9), we have (6.3) and, therefore, (6.2), by setting $a_{i,j} = \binom{i}{j} a_{i-j}$. Vice versa, if $\{a_n\}_n$ is an A. p. s., we have (6.2), and then (6.9) follows from Remark 6.1. □

So any polynomial $a_n(x)$ of an A. p. s. is determined by (6.9) by means of integration and relative constant.

Remark 6.2. The previous theorem, with the identity principle of polynomials, justifies the definition of an Appell-type matrix, as in Algorithm 2.2.1.

Remark 6.3. After Theorem 6.1, the sequence (6.2) is an A. p. s. as defined in [17].

Example 6.1. The monomial sequence $\{x^n\}_n$ is the Appell sequence with matrix $A = I$. It is the unique A. p. s. that is also b. p. s.

Other classical and nonclassical examples will be presented in the following.

It is possible to characterize an A. p. s. with a generating function [166, p. 1].

Let $\{a_n\}_n$ be an A. p. s. By Definition 6.1, there exists a numerical sequence $(a_i)_i$ with $a_0 \neq 0$ and $a_i = a_{i,0}$; so we can consider the formal power series

$$g(t) = \sum_{n=0}^{\infty} a_n \frac{t^n}{n!}. \tag{6.10}$$

Then we get the following:

Theorem 6.2 (Generating function). *A polynomial sequence $\{a_n\}_n$ is an A. p. s. if and only if there exists a formal power series $g(t)$ as in (6.10) such that*

$$g(t) \exp(xt) = \sum_{n=0}^{\infty} a_n(x) \frac{t^n}{n!}. \tag{6.11}$$

Proof. If $\{a_n\}_n$ is an A. p. s., by Definition 6.1, we have the numerical sequence $(a_n)_n$ with $a_n = a_{n,0}$, and there exists the formal power series as in (6.10). Then the result follows by the Cauchy product of series and Definition 6.1.

Vice versa, if (6.11) holds, we have the formal power series $g(t)$ as in (6.10), and, from the Cauchy product of series, we get

$$a_n(x) = \sum_{k=0}^{n} \binom{n}{k} a_{n-k} x^k.$$

Then the result follows from Definition 6.1. □

The function $g(t)\exp(xt)$ is called the *generating function* of A. p. s. $\{a_n\}$.

Corollary 6.1. *Let $\{a_n\}_n$ and $\{\widehat{a}_n\}_n$ be conjugate A. p. s. If $g(t)\exp(xt)$ is the generating function of $\{a_n\}_n$, then $\frac{1}{g(t)}\exp(xt)$ is the generating function of $\{\widehat{a}_n\}_n$, that is,*

$$\frac{1}{g(t)}\exp(xt) = \sum_{n=0}^{\infty} \widehat{a}_n(x)\frac{t^n}{n!}.$$

Proof. If

$$\frac{1}{g(t)} = \sum_{n=0}^{\infty} \widehat{a}_n \frac{t^n}{n!},$$

relation (6.7) holds, and then the result follows applying Cauchy product of series. □

Remark 6.4. The following *chain of equivalences* holds:

$$(6.2) \Leftrightarrow (6.3) \Leftrightarrow (6.9) \Leftrightarrow (6.11).$$

6.3 Recurrence relations

There are more interesting recurrence relations for A. p. s.

Let $\{a_n\}_n$ and $\{\widehat{a}_n\}_n$ be conjugate A. p. s. and $\{x^n\}_n$ the canonical sequence, we consider the following infinite vectors

$$\begin{aligned} A(x) &= [a_0(x), a_1(x), \dots, a_n(x), \dots]^T, \\ \widehat{A}(x) &= [\widehat{a}_0(x), \widehat{a}_1(x), \dots, \widehat{a}_n(x), \dots]^T, \\ X(x) &= [1, x, \dots, x^n, \dots]^T, \end{aligned} \tag{6.12}$$

and, consequently, for all $n \in \mathbb{N}$,

$$\begin{aligned} A_n(x) &= [a_0(x), a_1(x), \dots, a_n(x)]^T, \\ \widehat{A}_n(x) &= [\widehat{a}_0(x), \widehat{a}_1(x), \dots, \widehat{a}_n(x)]^T, \\ X_n(x) &= [1, x, \dots, x^n]^T. \end{aligned} \tag{6.13}$$

The vectors $A(x)$ and $\widehat{A}(x)$ are called Appell conjugate polynomial vectors [62, 214].

Lemma 6.1. *For the Appell vectors, we have*

$$\begin{aligned} A(x) &= AX(x), \quad \widehat{A}(x) = \widehat{A}X(x), \\ X(x) &= \widehat{A}A(x) = A\widehat{A}(x), \end{aligned} \tag{6.14}$$

and, consequently, for all $n \in \mathbb{N}$,

$$A_n(x) = A_nX_n(x), \quad \widehat{A}_n(x) = \widehat{A}_nX_n(x), \tag{6.15}$$

$$A_n(x) = A_n^2\widehat{A}_n(x), \quad \widehat{A}_n(x) = \widehat{A}_n^2 A_n(x), \tag{6.16}$$
$$X_n(x) = \widehat{A}_n A_n(x) = A_n\widehat{A}_n(x).$$

Proof. The proof follows from (6.2) and (6.6). □

Remark 6.5. We observe that the matrices $A^2, \widehat{A}^2$ are exponential Riordan matrices, in particular, Appell matrices. As a result,

$$A^2 = [g^2(x), x], \quad \widehat{A}^2 = \left[\frac{1}{g^2(x)}, x\right], \tag{6.17}$$

that is, if

$$A^2 := (\overline{a}_{i,k}), \quad \widehat{A}^2 := (\overline{\widehat{a}}_{i,k}), \tag{6.18}$$

we have

$$\overline{a}_{i,k} = \binom{i}{k}\sum_{j=0}^{i-k}\binom{i-k}{j} a_j a_{i-k-j}, \quad k = 0, \ldots, i, \quad i = 0, 1, \ldots, \tag{6.19}$$

$$\overline{\widehat{a}}_{i,k} = \binom{i}{k}\sum_{j=k}^{i}\binom{i-k}{j} \widehat{a}_j \widehat{a}_{i-k-j}, \quad k = 0, \ldots, i, \quad i = 0, 1, \ldots. \tag{6.20}$$

If we define

$$\overline{a}_n = \sum_{i=0}^{n}\binom{n}{i} a_i a_{n-i}, \tag{6.21}$$

$$\overline{\widehat{a}}_n = \sum_{i=0}^{n}\binom{n}{i} \widehat{a}_i \widehat{a}_{n-i}, \tag{6.22}$$

we have

$$\overline{a}_{i,k} = \binom{i}{k}\overline{a}_{i-k}, \tag{6.23}$$

$$\overline{\widehat{a}}_{i,k} = \binom{i}{k}\overline{\widehat{a}}_{i-k}. \tag{6.24}$$

Then we have

$$a_n(x) = \sum_{k=0}^{n} \overline{a}_{n,k}\widehat{a}_k(x), \quad \widehat{a}_n(x) = \sum_{k=0}^{n} \overline{\widehat{a}}_{n,k} a_k(x). \tag{6.25}$$

Denoting with $\{\overline{a}_n\}_n$ and $\{\overline{\widehat{a}}_n\}_n$ the A. p. s. related to the numerical sequences $(\overline{a}_n)$ and $(\overline{\widehat{a}}_n)$, we have that the A. p. s. $\{a_n\}$ is the umbral composition of the A. p. s. $\{\overline{a}_n\}_n$ and $\{\widehat{a}_n\}$, whereas the A. p. s. $\{\widehat{a}_n\}$ is the umbral composition of the A. p. s. $\{\overline{\widehat{a}}_n\}_n$ and $\{a_n\}_n$.

Now we have

Theorem 6.3 (First recurrence relation, [60]). *The polynomial sequences $\{a_n\}_n$, $\{\widehat{a}_n\}_n$ are conjugate A. p. s. if and only if the following recurrence relations hold:*

$$a_0(x) = \frac{1}{\widehat{a}_0}, \quad a_n(x) = \frac{1}{\widehat{a}_0}\left(x^n - \sum_{k=0}^{n-1}\binom{n}{k}\widehat{a}_{n-k}a_k(x)\right); \tag{6.26}$$

$$\widehat{a}_0(x) = \frac{1}{a_0}, \quad \widehat{a}_n(x) = \frac{1}{a_0}\left(x^n - \sum_{k=0}^{n-1}\binom{n}{k}a_{n-k}\widehat{a}_k(x)\right); \tag{6.27}$$

$$a_0(x) = \frac{1}{\overline{\widehat{a}}_0}, \quad a_n(x) = \frac{1}{\overline{\widehat{a}}_{n,n}}\left(\widehat{a}_n(x) - \sum_{k=0}^{n-1}\binom{n}{k}\overline{\widehat{a}}_{n,k}a_k(x)\right); \tag{6.28}$$

$$\widehat{a}_0(x) = \frac{1}{\overline{a}_0}, \quad \widehat{a}_n(x) = \frac{1}{\overline{a}_{n,n}}\left(a_n(x) - \sum_{k=0}^{n-1}\binom{n}{k}\overline{a}_{n,k}\widehat{a}_k(x)\right). \tag{6.29}$$

Proof. The result follows from (6.14), (6.15), (6.16). We observe that (6.26) and (6.27) also follow from (6.8). □

Corollary 6.2. *Let $\{a_n\}_n$ and $\{\widehat{a}_n\}_n$ be conjugate A. p. s. The sets $\{a_0, \dots, a_n\}$ and $\{\widehat{a}_0, \dots, \widehat{a}_n\}$ are bases for $\mathcal{P}_n$, and we have*

$$\sum_{k=0}^{n}\binom{n}{k}\widehat{a}_{n-k}a_k(x) = x^n = \sum_{k=0}^{n}\binom{n}{k}a_{n-k}\widehat{a}_k(x). \tag{6.30}$$

Proof. The result easily follows from (6.14) and (6.26). We observe that this result also follows from (6.8). □

Relation (6.30) allows us to write any polynomial in Appell basis.

Corollary 6.3. *Given the polynomial of degree $\le n$,*

$$p_n(x) = \sum_{k=0}^{n} c_{n,k}x^k, \tag{6.31}$$

we have

$$p_n(x) = \sum_{k=0}^{n}\widetilde{c}_{n,k}a_k(x) = \sum_{k=0}^{n}c^*_{n,k}\widehat{a}_k(x), \tag{6.32}$$

where

$$\begin{aligned}\widetilde{c}_{n,k} &= \sum_{j=0}^{n-k}\binom{k+j}{k}c_{n,k+j}\widehat{a}_k,\\ c^*_{n,k} &= \sum_{j=0}^{n-k}\binom{k+j}{k}c_{n,k+j}a_k,\end{aligned} \tag{6.33}$$

given $\{a_n\}_n$, $\{\widehat{a}_n\}_n$ conjugate A. p. s.

We can get another recurrence relation.

Theorem 6.4 (Second recurrence relation). *Let $\{a_n\}_n$ be a polynomial sequence. It is an A. p. s. if and only if there exists a numerical sequence (a_i) with $a_0 = 1$, so a formal power series $g(t) = \sum_{i=0}^{\infty} a_i \frac{t^i}{i!}$ such that the following relation holds:*

$$\begin{aligned} a_{n+1}(x) &= (x+b_0)a_n(x) + \sum_{k=0}^{n-1} \binom{n}{k} b_{n-k} a_k(x) \\ &= (x+b_0)a_n(x) + \sum_{k=1}^{n} \binom{n}{k} b_k a_{n-k}(x), \quad n = 0, 1, \dots, \end{aligned} \tag{6.34}$$

where $(b_i)_i$ are the coefficients of the following formal power series:

$$\frac{g'(t)}{g(t)} = \sum_{i=0}^{\infty} b_i \frac{t^i}{i!}. \tag{6.35}$$

Proof. Let $\{a_n\}_n$ be the A. p. s. with the numerical sequence $(a_i)_i$ with $a_0 = 1$. Then we have the generating function (6.11), that is,

$$g(t)\exp(xt) = \sum_{n=0}^{\infty} a_n(x) \frac{t^n}{n!},$$

with

$$g(t) = \sum_{i=0}^{\infty} a_i \frac{t^i}{i!}.$$

Partial derivation of the above identity with respect to the variable t gives

$$\left(\frac{g'(t)}{g(t)} + x\right) g(t)\exp(xt) = \sum_{n=0}^{\infty} a_{n+1}(x) \frac{t^n}{n!}.$$

Setting

$$\frac{g'(t)}{g(t)} = \sum_{i=0}^{\infty} b_i \frac{t^i}{i!},$$

after easy calculations, we have

$$\sum_{n=0}^{\infty} \left(\sum_{i=0}^{n} \binom{n}{i} b_{n-i} a_i(x) + x a_n(x) \right) \frac{t^n}{n!} = \sum_{n=0}^{\infty} a_{n+1}(x) \frac{t^n}{n!},$$

from which the result follows.

Vice versa, from (6.34), after easy calculation and keeping in mind relation (6.35) and the identity principle of polynomials, we get that $\{a_n\}$ is an A. p. s. as in Definition 6.1. □

Corollary 6.4. *For the numerical coefficients b_i, $i = 0, 1, \ldots$, of the above theorem, we have*

$$b_n = \sum_{k=0}^{n} \binom{n}{k} a_{k+1} \hat{a}_{n-k}, \quad n = 0, 1, \ldots, \tag{6.36}$$

where $\hat{a}_k$ are the coefficients of the power series $\frac{1}{g(t)}$, constructed as in (2.10).

Proof. The result follows from (6.35). □

Remark 6.6. The necessary condition of Theorem 6.4 is in [213], but it is proved with a more complex procedure.

Remark 6.7. We observe that if

$$\sum_{k=0}^{n-2} \binom{n}{k} b_{n-k} a_k(x) = 0 \quad \forall x \in \mathbb{R}, \quad \forall n \geq 2,$$

recurrence relation (6.34) becomes a three-term relation, and, consequently, in suitable hypothesis, the sequence $\{a_n\}$ is also orthogonal [90, 92, 191]. It is known [180] that among classical orthogonal polynomials only the Hermite sequence is—also—an A. p. s. We will consider this sequence afterwards.

6.4 Determinantal forms

For the Appell polynomial sequences, we can derive determinantal forms.

Theorem 6.5 (First determinantal form, [60]). *Let $\{a_n\}_n$ be a polynomial sequence. It is an A. p. s. if and only if there exists a numerical sequence $(a_n)_n$ with $a_0 = 1$ such that if $(\hat{a}_n)_n$ is the sequence defined as in (2.10), $\forall n \in \mathbb{N}$, we have*

$$\begin{aligned} a_0(x) &= \frac{1}{\hat{a}_0}, \\ a_n(x) &= \frac{(-1)^n}{(\hat{a}_0)^{n+1}} \begin{vmatrix} 1 & x & \cdots & x^{n-1} & x^n \\ \hat{a}_0 & \hat{a}_1 & \cdots & \hat{a}_{n-1} & \hat{a}_n \\ & \hat{a}_0 & \cdots & \binom{n-1}{1}\hat{a}_{n-2} & \binom{n}{1}\hat{a}_{n-1} \\ & & \ddots & & \vdots \\ & & & \hat{a}_0 & \binom{n}{n-1}\hat{a}_1 \end{vmatrix}. \end{aligned} \tag{6.37}$$

Proof. If $\{a_n\}_n$ is an A. p. s. with matrix $A = (a_{ij})$, we have (6.26), which is equivalent to an infinite triangular linear system. Solving by Cramer's rule, the system obtained from the first $n + 1$ equations in the unknowns $a_n(x)$, we have (6.37).

Vice versa, if (6.37) holds, by expanding the determinant with respect to the last column, after some calculations, we obtain (6.26), and, therefore, $\{a_n\}_n$ is an A. p. s. □

By symmetry, we have

$$\hat{a}_0 = \frac{1}{a_0},$$

$$\hat{a}_n(x) = \frac{(-1)^n}{(a_0)^{n+1}} \begin{vmatrix} 1 & x & \cdots & x^{n-1} & x^n \\ a_0 & a_1 & \cdots & a_{n-1} & a_n \\ & a_0 & \cdots & \binom{n-1}{1}a_{n-2} & \binom{n}{1}a_{n-1} \\ & & \ddots & & \vdots \\ & & & a_0 & \binom{n}{n-1}a_1 \end{vmatrix}. \tag{6.38}$$

We note that expanding (6.37) with respect to the first row, we obtain Definition 6.1, given the matrices A and $\widehat{A}$ being inverse of each other.

From the second recurrence relation, we have:

Theorem 6.6 (Second determinantal form). *Let $\{a_n\}_n$ be a polynomial sequence. It is an A. p. s. if and only if there exists a numerical sequence $(a_i)_i$ with $a_0 \neq 0$, and we have*

$$a_0(x) = 1,$$

$$a_{n+1}(x) = \begin{vmatrix} x+b_0 & -1 & 0 & & \cdots & & 0 \\ b_1 & x+b_0 & -1 & & \cdots & & 0 \\ b_2 & \binom{2}{1}b_1 & x+b_0 & -1 & \cdots & & 0 \\ \vdots & & & \ddots & & & \vdots \\ b_{n-1} & \binom{n-1}{1}b_{n-2} & \binom{n-1}{2}b_{n-3} & & \cdots & x+b_0 & -1 \\ b_n & \binom{n}{1}b_{n-1} & \binom{n}{2}b_{n-2} & & \cdots & \binom{n}{n-1}b_1 & x+b_0 \end{vmatrix}, \tag{6.39}$$

where $(b_i)_i$ is defined in (6.36).

Proof. The proof follows from Theorem 6.4. In fact, applying Cramer's rule for the solution of the first $n+1$ equations in the variables $a_1(x), \ldots, a_{n+1}(x)$, of infinite linear system (6.34), we have (6.39).

Vice versa, from (6.39), we have (6.34), with Laplace expansion of the determinant with respect to the last row. □

Corollary 6.5. *A polynomial sequence $\{a_n\}_n$ is an A. p. s. if and only if the polynomial $a_n(x)$ is the characteristic polynomial of a suitable Hessenberg matrix.*

Remark 6.8. The matrix $\widehat{R} := (\hat{r}_{i,j})_{i,j\geq 0}$, defined by

$$\begin{cases} \hat{r}_{i,j} = \binom{i}{j}b_{i-j}, & i > j, \\ \hat{r}_{i,i+1} = 1, & \\ \hat{r}_{i,j} = 0, & j > i+1. \end{cases}$$

where $(b_i)_i$ are as in (6.36), is the production matrix of $\widehat{A}$, conjugate of A. Hence, in the A. p. s. $\{a_n\}_n$ for any $n \in \mathbb{N}$, $a_n(x)$ is the characteristic polynomial of the production matrix of $\widehat{A}_n$.

Remark 6.9. The second recurrence relation and the second determinantal form for an A. p. s., only as a necessary condition, can be found in [213], but they are determined by a more general and complicated procedure.

From the previous theorems, we have

Theorem 6.7 (Equivalence theorem). *With the above notation, we have the following chain of equivalences:*

$$(6.2) \Leftrightarrow (6.3) \Leftrightarrow (6.9) \Leftrightarrow (6.11) \Leftrightarrow (6.26) \Leftrightarrow (6.34) \Leftrightarrow (6.37) \Leftrightarrow (6.39).$$

The recurrence relations generate differential equations for A. p. s. As an example, we will determine the differential equation generated by the first recurrence relation, but—with the same technique—we can get the differential equation that is obtained from the second recurrence relation, et cetera.

Theorem 6.8 (First differential equation, [62]). *Let $\{a_n\}_n$ be the A. p. s. with associated matrix $A = (a_{ij})$. Then $a_n(x)$ satisfies the differential equation*

$$\frac{\widehat{a}_n}{n!}y^{(n)}(x) + \frac{\widehat{a}_{n-1}}{(n-1)!}y^{(n-1)}(x) + \cdots + \widehat{a}_0 y(x) = x^n, \tag{6.40}$$

where $(\widehat{a}_i)_i$ is defined in Algorithm 2.2.2.

Proof. From Theorem 6.3, we have

$$a_{n+1}(x) = \frac{1}{\widehat{a}_0}\left(x^{n+1} - \sum_{k=0}^{n}\binom{n+1}{k+1}\widehat{a}_{k+1}a_{n-k}(x)\right), \tag{6.41}$$

but from (6.9), we find that

$$a'_{n+1}(x) = (n+1)a_n(x)$$

and, hence,

$$\forall k = 0, \dots, n, \quad a_{n-k}(x) = \frac{a_n^{(k)}(x)}{n(n-1)\dots(n-k+1)}. \tag{6.42}$$

By replacing $a_{n-k}(x)$, given in (6.42), in (6.41), we obtain

$$a_{n+1}(x) = \frac{1}{\widehat{a}_0}\left(x^{n+1} - (n+1)\sum_{k=0}^{n}\widehat{a}_{k+1}\frac{a_n^{(k)}(x)}{(k+1)!}\right). \tag{6.43}$$

Differentiating (6.41) and for (6.43), after some calculations, we get the result. □

Remark 6.10. Theorem 6.8 is in [102, Th. 2.1] and [215, Cor. 3.6], but with a very different and more complicated proof.

6.5 General properties

By elementary tools of matrix algebra, we can prove some general properties of Appell polynomials.

Let $\{p_n\}_n$ and $\{q_n\}_n$ be A. p. s. with matrix $P = (p_{ij})$ and $Q = (q_{ij})$, respectively. We indicate by $p_n(q_n(x))$ the umbral composition, and by $\lambda p_n(x) + \mu q_n(x)$, given $\lambda, \mu \in \mathbb{R}$, the linear combination of $p_n(x)$, $q_n(x)$. Then we have:

Theorem 6.9 (Algebraic structure, [62]). *The following statements hold:*

1. *Let $\mathcal{A}_n = \{\{p_n\}_n \mid \{p_n\}_n \text{is an A. p. s.}\}$. Then $\mathcal{A}$ with the umbral composition*

$$p_n \circ q_n = p_n(q_n(x))$$

   is a group, in which
   a. *the identity is $\{x^n\}_n$,*
   b. *the inverse of $p_n(x)$ is its conjugate $\widehat{p}_n(x)$.*
2. *The linear combination $\lambda p_n(x) + \mu q_n(x)$, $\lambda, \mu \in \mathbb{R}$, with $p_n(x)$, $q_n(x)$ A.p.s., is an Appell sequence.*

Proof.
1. (a) is obvious, whereas for (b), we observe that for the conjugate sequence there is (6.8).
2. The result follows from the first determinantal form by applying the property of linearity of the determinant [60, 62]. □

Remark 6.11. The algebraic structure $\{\mathcal{A}_n, \circ, +, \cdot\}$ is an algebra over $\mathbb{K}$.

Theorem 6.10 (Connection constants, [62]). *Let $A(x)$ and $P(x)$ be Appell vectors with matrix $A = (a_{ij})$ and $P = (p_{ij})$, respectively. Then*

$$\forall n \in \mathbb{N}, \quad A_n(x) = C_n P_n(x), \tag{6.44}$$

where C_n is the matrix of entries

$$(C)_{ij} = \begin{cases} \binom{i}{j} c_{i-j}, & i \geq j, \\ 0 & \text{otherwise}, \end{cases} \qquad i = 0, 1, \dots, n, \tag{6.45}$$

with

$$c_n = \sum_{k=0}^{n} \binom{n}{k} a_{n-k} \widehat{p}_k, \quad n = 0, 1, \dots.$$

Proof. The proof follows after easy calculations from (6.14) and (6.15). □

Theorem 6.11 (Inverse relation, [62]). *Let $\{a_n\}_n$ and $\{\widehat{a}_n\}_n$ be conjugate A. p. s., with sequences (a_i), $(\widehat{a}_i)$. Then from*

$$y_n = \sum_{k=0}^{n} \binom{n}{k} \widehat{a}_{n-k} x_k \tag{6.46}$$

it follows

$$x_n = \sum_{k=0}^{n} \binom{n}{k} a_{n-k} y_k, \tag{6.47}$$

that is, (6.46) *and* (6.47) *are of inverse relations.*

Proof. Putting $\overline{y}_n = [y_0, \dots, y_n]^T$ and $\overline{x}_n = [x_0, \dots, x_n]^T$, (6.46) and (6.47) are equivalent to

$$\begin{cases} \overline{y}_n = \widehat{A}_n \overline{x}_n, \\ \overline{x}_n = A_n \overline{y}_n. \end{cases}$$

But $A_n \widehat{A}_n = I$ from (2.25); hence, (6.46) and (6.47) are of inverse relations. □

Theorem 6.12 (Inverse relation between two Appell polynomial sequences, [62]). *Let $P_n(x)$ and $Q_n(x)$ be Appell vectors, with matrix $P = (p_{ij})$ and $Q = (q_{ij})$, respectively. Then, according to the previous notations and hypothesis, the following are of inverse relations:*

$$\forall n \in \mathbb{N} : \begin{cases} P_n(x) = \overline{P}_n Q_n(x), \\ Q_n(x) = \overline{Q}_n P_n(x), \end{cases} \tag{6.48}$$

with

$$(\overline{P})_{ij} = \begin{cases} \binom{i}{j} \overline{p}_{i-j}, & i \geq j, \\ 0 & \text{otherwise}, \end{cases} \qquad \text{where} \quad \overline{p}_n = \sum_{k=0}^{n} \binom{n}{k} p_{n-k}(0) \widehat{q}_k, \tag{6.49}$$

$$(\overline{Q})_{ij} = \begin{cases} \binom{i}{j} \overline{q}_{i-j}, & i \geq j, \\ 0 & \text{otherwise}, \end{cases} \qquad \text{where} \quad \overline{q}_n = \sum_{k=0}^{n} \binom{n}{k} q_{n-k}(0) \widehat{p}_k. \tag{6.50}$$

Proof. The proof follows from Theorem 6.10, observing that

$$\sum_{k=0}^{n} \binom{n}{k} \overline{p}_{n-k} \overline{q}_k = \delta_{n,0}, \quad n = 0, 1, \dots, \tag{6.51}$$

and, therefore,

$$\overline{P}_n \overline{Q}_n = I_{n+1}.$$

□

Theorem 6.13 (Binomial identity, [60]). *Let $\{a_n\}_n$ be a polynomial sequence. It is an A. p. s. if and only if*

$$a_n(x+y) = \sum_{i=0}^{n} \binom{n}{i} a_i(x) y^{n-i}, \quad n = 0, 1, \ldots, \quad \forall x, y \in \mathbb{R}. \tag{6.52}$$

Proof. If $\{a_n\}$ is an A. p. s. the proof follows by the known binomial identity and from the first determinantal form.

Vice versa, from (6.52) for $x = 0$, we have

$$a_n(y) = \sum_{i=0}^{n} \binom{n}{i} a_i(0) y^{n-i} = \sum_{i=0}^{n} \binom{n}{i} a_{n-i}(0) y^i,$$

that is relation (6.4). So $\{a_n\}_n$ is an A. p. s. □

Corollary 6.6. *From the Binomial identity of an A. p. s., we get the known combinatorial identity* [100, *p. 119*]

$$\sum_{k=j}^{n} (-1)^k \binom{n}{k} \binom{k}{j} = (-1)^n \delta_{n,j}. \tag{6.53}$$

Proof. If we set $y = -x$ in (6.52), we get, after easy calculations,

$$a_n(0) = a_n + \sum_{j=0}^{n-1} a_j x^{n-j} \sum_{k=j}^{n} (-1)^{n-j} \binom{n}{j} \binom{j}{k}.$$

Then the result follows using the identity principle of polynomials. □

Remark 6.12. From Theorem 6.13, the *chain of equivalences* of definitions of A. p. s. can be extended. In fact, we have

$$(6.2) \Leftrightarrow (6.3) \Leftrightarrow (6.9) \Leftrightarrow (6.11) \Leftrightarrow (6.26) \Leftrightarrow (6.34) \Leftrightarrow (6.37) \Leftrightarrow (6.39) \Leftrightarrow (6.52).$$

Theorem 6.14 (Generalized Appell identity, [62]). *Let $\{a_n\}$ and $\{b_n\}$ be A. p. s. with matrices $A = (a_{ij})$ and $B = (b_{ij})$. If $\{c_n\}_n$ is the A. p. s. with matrix $C = (c_{i,j})$, where*

$$c_{i,j} = \begin{cases} \binom{i}{j} c_{i-j}, & i = 0, 1, \ldots, \quad j = 0, \ldots, i, \\ 0, & i < j, \end{cases} \tag{6.54}$$

and

$$c_0 = \frac{1}{c_0(0)}, \quad c_i = -\frac{1}{c_0(0)} \sum_{k=1}^{i} \binom{i}{k} c_{i-k} c_k(0), \quad i > 0, \tag{6.55}$$

with

$$c_k(0) = \sum_{j=0}^{k} \binom{k}{j} b_{k-j}(0) a_j(0), \quad k \geq 0,$$

then we have

$$c_n(x+y) = \sum_{k=0}^{n} \binom{n}{k} b_{n-k}(y) a_k(x). \tag{6.56}$$

Proof. The result follows from (6.2), (6.9), and previous results, putting alternately $x = 0$ and $y = 0$ in the (6.52). □

Theorem 6.15 (Combinatorial identities, [62]). *Let $\{a_n\}_n$ and $\{c_n\}_n$ be A.p.s with matrices $A = (a_{ij})$ and $C = (c_{ij})$, respectively. The following relations hold:*

$$\sum_{k=0}^{n} \binom{n}{k} a_k(x) c_{n-k}(-x) = \sum_{k=0}^{n} \binom{n}{k} a_k(0) c_{n-k}(0), \tag{6.57}$$

$$\sum_{k=0}^{n} \binom{n}{k} a_k(x) c_{n-k}(z) = \sum_{k=0}^{n} \binom{n}{k} a_k(x+z) c_{n-k}(0). \tag{6.58}$$

Proof. The results follow by the Theorem 6.14 after easy calculations. □

Theorem 6.16 (Forward difference, [60, 62]). *If $\{a_n\}_n$ is the A. p. s. for matrix $A = (a_{ij})$, we have*

$$\Delta a_n(x) := a_n(x+1) - a_n(x) = \sum_{i=0}^{n-1} \binom{n}{i} a_i(x), \quad n = 0, 1, \ldots. \tag{6.59}$$

Proof. The result follows from (6.52) for $y = 1$. □

Theorem 6.17 (Multiplication theorem, [62]). *Let $A_n(x)$, $n \in \mathbb{N}$, be the Appell vector with matrix $A_n = (a_{ij})$. The following identities hold:*

$$\begin{cases} A_n(mx) = E(x) A_n(x), & m = 1, 2, \ldots, \\ A_n(mx) = A_n^{-1} D X_n(x), & m = 1, 2, \ldots, \end{cases} \tag{6.60}$$

where

$$(E(x))_{ij} = \begin{cases} \binom{i}{j} (m-1)^{i-j} x^{i-j}, & i \geq 0, j < i, \\ 0, & i < j, \quad i, j = 1, \ldots, n, \end{cases}$$

$$D = \operatorname{diag}[1, m, \ldots, m^n].$$

Proof. The first of (6.60) follows from (6.56), putting $y = (m-1)x$, whereas the second follows from (6.15)) using previous results. □

Now we can consider the symmetry of A. p. s.

Theorem 6.18 (Symmetry, [60]). *For A. p. s. $\{a_n\}_n$, the following relation holds:*

$$(a_n(h-x) = (-1)^n a_n(x)) \Leftrightarrow (a_n(h) = (-1)^n a_n(0)), \quad h \in \mathbb{R}, \quad n = 0, 1, \ldots. \tag{6.61}$$

Proof. The sufficient condition follows from the hypothesis with $x = 0$.

For the necessary condition, using (6.52), we find

$$a_n(h-x) = \sum_{i=0}^{n} \binom{n}{i} a_i(h)(-x)^{n-i} = (-1)^n \sum_{i=0}^{n} \binom{n}{i} a_i(h)(-1)^i (x)^{n-i}$$
$$= (-1)^n \sum_{i=0}^{n} \binom{n}{i} a_{n-i}(h)(-1)^{n-i}(x)^i.$$

Therefore, using the hypothesis and (6.4), we have

$$a_n(h-x) = (-1)^n \sum_{i=0}^{n} \binom{n}{i} a_{n-i}(0)(x)^i = (-1)^n a_n(x).$$ □

Lemma 6.2 ([60]). *For the numerical sequences* $(a_n)_{n\in\mathbb{N}}$ *and* $(\widehat{a})_{n\in\mathbb{N}}$*, which generate the A. p. s.* $\{a_n\}_n$ *and* $\{\widehat{a}_n\}_n$*, we have*

$$(a_{2n+1} = 0) \Leftrightarrow (\widehat{a}_{2n+1} = 0). \tag{6.62}$$

Proof. From (2.13), we have

$$\widehat{a}_0 = \frac{1}{a_0},$$
$$\widehat{a}_n = -\frac{1}{a_0} \sum_{k=1}^{n-1} \binom{n}{k} a_k \widehat{a}_{n-k},$$

from which

$$\widehat{a}_1 = -\frac{1}{a_0} a_1 \widehat{a}_0,$$
$$\widehat{a}_{2n+1} = -\frac{1}{a_0} \binom{2n+1}{1} a_1 \widehat{a}_{2n} - \frac{1}{a_0} \left(\sum_{k=1}^{n} \binom{2n+1}{2k} a_{2k} \widehat{a}_{2(n-k)+1} + \sum_{k=1}^{n} \binom{2n+1}{2k+1} a_{2k+1} \widehat{a}_{2(n-k)} \right),$$

and

$$\forall \quad n = 0, 1, \ldots, \quad a_{2n+1} = 0 \quad \Rightarrow \quad \widehat{a}_{2n+1} = 0.$$

Likewise, we obtain the opposite. □

Theorem 6.19 ([60]). *For A. p. s.* $\{a_n\}$*, the following relation holds:*

$$(a_n(-x) = (-1)^n a_n(x)) \Leftrightarrow (a_{2n+1} = 0) \Leftrightarrow (\widehat{a}_{2n+1} = 0). \tag{6.63}$$

Proof. From Theorem 6.18 with $h = 0$ and Lemma 6.2, we find

$$(a_n(-x) = (-1)^n a_n(x)) \Leftrightarrow (a_n(0) = (-1)^n a_n(0))$$
$$\Leftrightarrow (a_{2n+1} = 0) \Leftrightarrow (\widehat{a}_{2n+1} = 0).$$ □

Theorem 6.20 (Integration, [60]). *For each $n \geq 1$, it is true that*

$$\int_0^x a_n(t)dt = \frac{1}{n+1}[a_{n+1}(x) - a_{n+1}(0)] \tag{6.64}$$

and

$$\int_0^1 a_n(x)dx = \frac{1}{n+1}\sum_{i=0}^{n}\binom{n+1}{i}a_i(0). \tag{6.65}$$

Proof. Relations (6.64) and (6.65) follow from (6.9) and Theorem 6.13 for $x = 0$ and $y = 1$. □

6.6 Localization of zeros

Theorem 6.21. *Let $\{a_n\}_n$ be an A.p.s. with associated matrix A, $\hat{A}$ its conjugate, and $\widehat{R} := (\widehat{r}_{ij})_{i,j\geq 0}$, the production matrix of $\widehat{A}$, as in Remark 6.1. Then the zeros of $a_n(x)$ lie in the circle of center $\frac{a_1}{a_0}$ and radius r, with*

$$r = \max_{0\leq i\leq n}\sum_{\substack{j=0\\ i\neq j}}^{i+1}|\widehat{r}_{i,j}|. \tag{6.66}$$

Proof. By Theorem 6.6 the zeros of $a_n(x)$ are the eigenvalues of the matrix $\widehat{R} = (\widehat{r}_{i,j})$, with $\widehat{r}_{i,j} = \binom{i}{j}b_{i-j}$ and $(b_i)_i$, given in (6.36). Then by Gersghorin's theorem [95], we have the result. □

Theorem 6.22. *For the zeros of Appell polynomials $a_n(x)$, the following relations are true:*

1. *If $a_n(\alpha) = 0$, $\alpha \in \mathbb{R}$, then*

$$a_n'(\alpha)a_{n-1}(\alpha) > 0. \tag{6.67}$$

2. *If $a_n(\alpha_1) = 0 = a_n(\alpha_2)$, with $\alpha_1, \alpha_2 \in \mathbb{R}$, $\alpha_1 \neq \alpha_2$, there exists $\beta \in \mathbb{R}$ such that*

$$\alpha_1 < \beta < \alpha_2 \quad \textit{and} \quad a_{n-1}(\beta) = 0. \tag{6.68}$$

3. *If the zeros of $a_n(x)$ are real and distinct, then the zeros of $a_n(x)$ and $a_{n+1}(x)$ are separated, and in a zero $\alpha_{i,n}$ of $a_n(x)$, it results that*

$$a_{n-1}(\alpha_{i,n})a_{n+1}(\alpha_{i,n}) < 0. \tag{6.69}$$

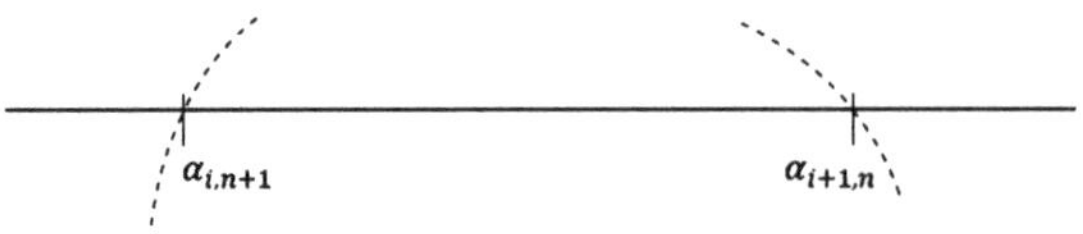

Figure 6.1: Sketch of the graphic of $P(x)$.

Proof.
1. Inequality (6.67) follows from (6.9).
2. Relation (6.68) follows from (6.9) and some known elementary property of real continuous functions.
3. Denoting by $\alpha_{1,n} < \alpha_{2,n} < \cdots < \alpha_{n,n}$ the zeros of $a_n(x)$, from (6.68) it follows that

$$\alpha_{1,n+1} < \alpha_{1,n} < \alpha_{2,n+1} < \alpha_{2,n} < \cdots, \tag{6.70}$$

that is, the zeros of $a_n(x)$ and $a_{n+1}(x)$, are separated. Furthermore, we can consider the polynomial [51]

$$P(x) = a_n(x)a_{n+1}(x),$$

and the result is

$$P'(x) = na_{n-1}(x)a_{n+1}(x) + (n+1)a_n^2(x),$$

from which

$$P'(\alpha_{i,n}) = na_{n-1}(\alpha_{i,n})a_{n+1}(\alpha_{i,n}) < 0, \tag{6.71}$$
$$P'(\alpha_{i,n+1}) = (n+1)a_n^2(\alpha_{i,n+1}) > 0. \tag{6.72}$$

From (6.70), (6.71), (6.72), it follows that $P(x)$ is increasing at $\alpha_{i,n+1}$ and decreasing in $\alpha_{i+1,n}$ (Figure 6.1); consequently, the result holds. □

Theorem 6.23. *Let $\{a_n\}_n$ be an A. p. s. Then the sequence*

$$a_n(x), a_{n-1}(x), \ldots, a_0(x) \quad \forall n \in \mathbb{N}$$

is a Sturm sequence [108].

Proof. By setting

$$q_k(x) = a_{n-k}(x), \quad k = 0, \ldots, n, \tag{6.73}$$

we have a Sturm sequence [108, p. 126] after the application of the previous theorem. □

Corollary 6.7. *If we denote by $V(x)$ the number of changes of signs of the values of the Sturm sequence* (6.73) *on the boundary of the interval $[a, b] \subset \mathbb{R}$, the difference $V(a) - V(b)$ is the number of zeros of $a_n(x)$ in the interval (a, b).*

Proof. The result follows using Sturm theorem [108, p. 126]. □

6.7 Operational matrices

In recent years, operational matrices have been employed for solving many engineering and physical problems, such as dynamical systems [204], optimal central systems [162], and robotic systems [26]. Furthermore, they are used in several areas of numerical analysis, such as differential equations [216] and integral equations [195].

Now we consider the operational matrices of Appell sequences.

Let $A_n(x)$ be the Appell vector. The matrices $\mathcal{D}$ and $\mathcal{I}$ are, respectively, called operational matrices of derivations and integration if and only if

$$\frac{d}{dx}A_n(x) = \mathcal{D}A_n(x), \tag{6.74}$$

$$\int_0^x A_n(x)dx \simeq \mathcal{I}A_n(x). \tag{6.75}$$

Then we have:

Theorem 6.24. *Let $A_n(x)$ be the Appell vector. $\mathcal{D}$ is the operational matrix of derivatives if and only if $\mathcal{D} = (D_{i,j})_{i,j=0,\ldots,n}$ with*

$$D_{i,j} = \begin{cases} i, & i = j+1, \\ 0 & \text{otherwise}, \end{cases} \qquad i,j = 0,\ldots,n. \tag{6.76}$$

Proof. The proof follows from (6.9) and (6.74). □

Remark 6.13. From (6.76), the matrix $\mathcal{D}$ is

$$\begin{bmatrix} 0 & 0 & \cdots & 0 & 0 \\ 1 & 0 & \cdots & 0 & 0 \\ 0 & 2 & \ddots & 0 & 0 \\ \vdots & & \ddots & & \vdots \\ 0 & \cdots & & n & 0 \end{bmatrix}. \tag{6.77}$$

This matrix in [4] is called *creation matrix* and the following theorem holds:

Theorem 6.25 ([4]). *If $G(x,t) = f(t)\exp(xt)$ is the generating function for an A. p. s. $\{a_n\}_n$ with matrix A, the following relation holds:*

$$A = f(\mathcal{D}). \tag{6.78}$$

For operational matrix of integration [171], we have

$$\int_0^x A_n(x)dx = \begin{bmatrix} \int_0^x a_0(t)dt \\ \int_0^x a_1(t)dt \\ \vdots \\ \int_0^x a_n(t)dt \end{bmatrix} = \begin{bmatrix} a_1(x) - a_1(0) \\ \frac{a_2(x)-a_2(0)}{2} \\ \vdots \\ \frac{a_{n+1}(x)-a_{n+1}(0)}{n+1} \end{bmatrix}. \tag{6.79}$$

Developing each component, we have

$$\begin{aligned}\frac{a_i(x)-a_i(0)}{i} &= \frac{1}{i}\sum_{j=0}^{i}\binom{i}{j}a_j(0)x^{i-j}-\frac{1}{i}a_i(0) \\ &= \frac{1}{i}\binom{i}{i-1}a_{i-1}(0)x+\frac{1}{i}\binom{i}{i-2}a_{i-2}(0)x^2+\cdots+\frac{1}{i}\binom{i}{0}a_0(0)x^i \\ &= \left[0,\frac{1}{i}\binom{i}{i-1}a_{i-1}(0),\ldots,\frac{1}{i}\binom{i}{0}a_0(0),\underbrace{0,\ldots,0}_{n-i}\right]X_n(x) \\ &= U_iX_n(x)=U_i\widehat{A}_nA_n(x),\quad i=1,\ldots,n,\end{aligned}\tag{6.80}$$

where

$$U_i=\left[0,\frac{1}{i}\binom{i}{i-1}a_{i-1}(0),\ldots,\frac{1}{i}\binom{i}{0}a_0(0),\underbrace{0,\ldots,0}_{n-i}\right].$$

So we just need to approximate the last component

$$\frac{a_{n+1}(x)-a_{n+1}(0)}{n+1}\simeq\sum_{i=0}^{n}w_ia_i(x)=W^TA_n(x)\tag{6.81}$$

with $W=[w_0,\ldots,w_n]$. Then by substituting (6.81) and (6.80) into (6.79) and from (6.15), we have

$$\int_0^x A_n(t)dt=\begin{bmatrix}U_1\widehat{A}_nA_n(x)\\ \vdots\\ U_n\widehat{A}_nA_n(x)\\ W^TA_n\widehat{A}_nA_n(x)\end{bmatrix}=\begin{bmatrix}U_1\\ \vdots\\ U_n\\ W^TA_n\end{bmatrix}\widehat{A}_nA_n(x)=\overline{U}\widehat{A}_nA_n(x)\tag{6.82}$$

with

$$\overline{U}=[U_1,\ldots,U_n,W^TA_n]^T.\tag{6.83}$$

From (6.82), we finally have

$$\mathcal{I}=\overline{U}\widehat{A}_n\tag{6.84}$$

6.8 Relationship with linear functionals

Let L be a linear functional defined on $\mathcal{P}$, the set of polynomials. If

$$L(1)\neq 0,\quad \hat{a}_i=L(x^i),\quad i=0,1,\ldots,\tag{6.85}$$

we have

Definition 6.2 ([62]). The polynomial sequence $\{a_n\}_n$—defined in (6.26), where $(\widehat{a}_n)_n$ is considered as in (6.85)—is called the Appell sequence related to a linear functional L, and, where necessary, it will be indicated as $\{a_{L,n}\}_n$.

Theorem 6.26 ([62]). *Let $\{a_n\}_n$ be an A. p. s. with sequences $(a_i)_i$ and $(\widehat{a}_i)_i$ as in (6.5). Then—if $G(x,t) = g(t)\exp(xt)$ is the generating function of $\{a_n\}_n$, and L is the linear functional defined in (6.85)—we have*

$$G(x,t) = \frac{\exp(xt)}{L_x(\exp(xt))}, \tag{6.86}$$

where L_x means that the functional L is applied to the argument as a function of x.

Proof. If the generating function is

$$G(x,t) = g(t)\exp(xt)$$

with

$$g(t) = \sum_{n=0}^{\infty} a_n \frac{t^n}{n!},$$

we have

$$\frac{1}{g(t)} = \sum_{n=0}^{\infty} \widehat{a}_n \frac{t^n}{n!}$$

with $\{\widehat{a}_n\}_n$ as in (6.5). Then we have

$$G(x,t) = \frac{\exp(xt)}{\frac{1}{g(t)}} = \frac{\exp(xt)}{\sum_{n=0}^{\infty} \widehat{a}_n \frac{t^n}{n!}} = \frac{\exp(xt)}{\sum_{n=0}^{\infty} L(x^n)\frac{t^n}{n!}} = \frac{\exp(xt)}{L_x(\exp(xt))}. \qquad \square$$

We note that Theorem 6.26, combined with Definition 6.2, can be considered as a generalization of probabilistic approach to Appell polynomials considered in [192].

Remark 6.14. We observe explicitly that if $\widehat{a}_i = L(x^i)$, for the A. p. s. $\{a_n\}_n$ generated by $(a_i)_i$ as in (6.5), we have the generating function

$$G(x,t) = \frac{\exp(xt)}{L_x(\exp(xt))}.$$

Now we define the $n+1$ linear functionals on $\mathcal{P}$ by

$$L_0(x^i) = L(x^i), \quad L_j(x^i) = L(D^j(x^i)) \quad \text{for } j \le i, \quad i = 1, \dots, n, \tag{6.87}$$

Then we have:

Theorem 6.27 ([62, 67]). *For the A. p. s. $\{a_{L,n}\}$, the result is*

$$L_i(a_{L,n}(x)) = n!\delta_{i,n}, \quad i = 0, \dots, n.$$

Proof. The proof follows from Theorem 6.5. □

Corollary 6.8. *The Appell sequence $\{a_{L,n}(x)\}$ is the solution of the general linear interpolation problem*

$$L_i(p_n(x)) = n!\delta_{i,n}, \quad i = 0, \dots, n, \tag{6.88}$$

where L_i is defined in (6.87).

Proof. Relation (6.88) follows using the known theorems on linear interpolation [76]. □

Remark 6.15. We note that (6.88) is equivalent to Theorem 2.3.1 of [167] for Appell polynomials.

Theorem 6.28 (Representation theorem, [67]). *With the previous hypothesis and notations, for any $P_n(x) \in \mathcal{P}_n$, we have*

$$P_n(x) = \sum_{k=0}^{n} \frac{L(P_n^{(k)}(x))}{k!} a_{L,k}(x). \tag{6.89}$$

Proof. Relation (6.89) follows from Theorem 6.27. □

The polynomial (6.89) is a natural generalization of the classic Taylor polynomials.

6.9 Generalization of Appell polynomials

The Appell polynomial sequences were subject to more generalizations (see [116, 117, 194] and reference therein), mostly depending on the type of applications, where they could be advantageously used, including multivariate, commutative, and noncommutative settings. The set of Sheffer A-zero-type polynomials [178, 179] can be considered as the first generalization. Now we only outline the so called q-Appell polynomials ([12, 130, 196] and reference in therein), given the q-calculus-based [114].

Let q be an arbitrary real or complex number, and let us define the q-derivative of a function $f(x)$ by means of

$$D_q f(x) = \frac{f(qx) - f(x)}{(q-1)x}, \quad q \neq 1, \tag{6.90}$$

that is, a generalization of the differential operator d/dx.

Then the q-Appell polynomials are the polynomial sets $\{a_n^q\}$, which satisfy

$$\begin{cases} D_q a_n^q(x) = [n]_q a_{n-1}^q(x), \\ a_0^q(x) = k \neq 0, \end{cases} \quad n = 1, 2, \dots, \tag{6.91}$$

where

$$[n]_q = (q^n - 1)/(q - 1), \quad q \neq 1. \tag{6.92}$$

These polynomials were first introduced in [178], where they were called q-harmonic (see also [12, 130] and references therein).

We note that when $q \longrightarrow 1$, (6.91) becomes

$$\frac{d}{dx} a_n^q(x) = n a_{n-1}^q(x), \tag{6.93}$$

so that we may think of a q-Appell polynomial set as a generalization of Appell polynomials. Recently, an algebraic theory for q-Appell polynomials [178] has been developed, but a similar matrix theory of q-Appell polynomials is possible, defining the infinite lower triangular matrix $A \equiv (a_{i,j})$ with

$$a_{i,j} = \begin{bmatrix} i \\ j \end{bmatrix}_q a_{i-j}, \tag{6.94}$$

where $(a_i)_i$ is a numerical sequence with $a_0 \neq 0$, and

$$\begin{bmatrix} i \\ j \end{bmatrix}_q = \frac{[i]_q!}{[j]_q![i-j]_q!};$$

moreover,

$$[i]_q! = \prod_{k=1}^{n} [k]_q, \quad [0]_q! = 1.$$

6.10 Summary

The Appell polynomial sequences have been introduced by means of matrix calculus.

Let $(a_i)_i$ with $a_0 \neq 0$ be a numerical sequence. Then we can define an infinite, lower triangular matrix of Appell type.

This matrix generates the Appell polynomial sequence $\{a_n\}_n$. Then properties and characterizations are given. In particular recurrence, differential relations, and determinantal forms have been proved. The second recurrence relation allows an interesting link with orthogonal polynomials by means of Favard's theorem.

A link with linear functional, hence with Rota et al.'s theory, is given. Finally, new and interesting properties on zeros are sketched, and some possible generalizations are mentioned.

7 Application to linear interpolation and approximation theory

Abstract: The Appell polynomial sequences (A. p. s.) have application in approximation theory of regular functions. In the following, we focus on the linear interpolation and operators approximation theory.

7.1 Appell interpolation

Let X be the linear space of real continuous functions defined in the interval $[0, 1]$ and with continuous derivatives of all necessary orders. Let L be a linear functional on X such that $L(1) \neq 0$ and $\{a_{L,n}\}_n$ the A. p. s. relative to linear functional L such that it is the A. p. s. defined by $(a_i)_i$ with a_i given by (6.7), where $\widehat{a}_i = L(x^i)$, $i = 0, 1, \ldots$.

Then we have:

Theorem 7.1 ([67]). *For any $f \in X$, the polynomial*

$$P_{L,n}[f](x) = \sum_{i=0}^{n} \frac{L(f^{(i)})}{i!} a_{L,i}(x) \tag{7.1}$$

is the unique polynomial of degree $\leq n$ such that

$$L(P_{L,n}[f]^{(i)}) = L(f^{(i)}), \quad i = 0, \ldots, n. \tag{7.2}$$

Proof. Let

$$L_0(x^i) = L(x^i), \quad i = 0, 1, \ldots, n,$$
$$L_j(x^i) = L(D^j x^i), \quad j = 1, \ldots, i.$$

The result follows from Theorems 6.27 and 6.28. □

Definition 7.1 ([67]). The polynomial (7.1) is called the *Appell interpolant polynomial* for f related to a functional L.

Now it is interesting to consider the estimation of the remainder

$$R_{L,n}[f](x) = f(x) - P_{L,n}[f](x). \tag{7.3}$$

Remark 7.1 ([67]). For $f \in \mathcal{P}_n$, we have

$$R_{L,n}[f](x) = 0 \quad \text{and} \quad R_{L,n}[x^{n+1}] \neq 0 \quad \forall x \in [0, 1], \tag{7.4}$$

that is, the polynomial operator (7.1) is exact on $\mathcal{P}_n$.

https://doi.org/10.1515/9783110652925-007

For a fixed x, we may consider the remainder $R_{L,n}[f](x)$ as a linear functional with argument f, which vanishes on all the elements of $\mathcal{P}_n$. From Peano's theorem [76] if a linear functional has this property, it must have a simple representation in terms of $f^{(n+1)}$.

Therefore, we have

Theorem 7.2 ([67]). *Let $f \in C^{n+1}[a,b]$. The following relation holds:*

$$R_{L,n}[f](x) = \frac{1}{n!}\int_0^1 K_n(x,t) f^{(n+1)}(t)dt \quad \forall x \in [0,1], \tag{7.5}$$

where

$$K_n(x,t) = (x-t)_+^n - \sum_{i=0}^{n} \binom{n}{i} L((x-t)_+^{n-i}) a_{L,i}(x). \tag{7.6}$$

Proof. After some calculation, the result follows using Remark 7.1 and Peano's theorem. □

Remark 7.2 ([67]). If $f^{(n+1)} \in C^p[0,1]$ and $K_n(x,t) \in C^q[0,1]$ with $\frac{1}{p} + \frac{1}{q} = 1$, then we apply the well-known Hölder inequality

$$|R_{L,n}[f](x)| \le \frac{1}{n!}\left(\int_0^1 |K_n(x,t)|^q dt\right)^{\frac{1}{q}} \left(\int_0^1 |f^{(k+1)}(t)|^p dt\right)^{\frac{1}{p}}. \tag{7.7}$$

Two important cases are:

1. $p = q = 2$
   In this case,

$$|R_{L,n}[f](x)| \le \sigma_n \|f\|, \tag{7.8}$$

   with

$$(\sigma_n)^2 = \left(\frac{1}{n!}\right)^2 \int_0^1 (K_n(x,t))^2 dt \quad \|f\|^2 = \int_0^1 (f^{(n+1)}(t))^2 dt.$$

2. $q = 1, p = \infty$
   We have

$$|R_{L,n}[f](x)| \le \frac{1}{n!} M_{n+1} \int_0^1 |K_n(x,t)| dt, \tag{7.9}$$

   where

$$M_{n+1} = \sup_{0 \le x \le 1} |f^{(n+1)}(x)|.$$

A further polynomial operator can be determined as follows:
For any fixed $z \in [0,1]$, we consider the polynomial

$$\overline{P}_{L,n}[f](x) = f(z) + P_{L,n}[f](x) - P_{L,n}[f](z) = f(z) + \sum_{i=0}^{n} \frac{L(f^{(i)})}{i!}(a_{L,i}(x) - a_{L,i}(z)). \tag{7.10}$$

Then we have the following:

Theorem 7.3 ([67]). *The polynomial $\overline{P}_{L,n}[f](x)$ is an approximating polynomial of order n for $f(x)$, that is,*

$$\forall x \in [0,1], \quad f(x) = \overline{P}_{L,n}[f](x) + \overline{R}_{L,n}[f](x), \tag{7.11}$$

where the remainder $\overline{R}_{L,n}[f](x)$ satisfies

$$\overline{R}_{L,n}[t^i](x) = 0, \quad i = 0,\ldots,n, \quad \overline{R}_{L,n}[t^{n+1}](x) \neq 0. \tag{7.12}$$

Proof. For all $x \in [0,1]$ and for any fixed $z \in [0,1]$, from (7.3), we obtain

$$f(x) - f(z) = P_{L,n}[f](x) - P_{L,n}[f](z) + R_{L,n}[f](x) - R_{L,n}[f](z),$$

from which we get (7.11) and (7.12). The exactness of polynomials $\overline{P}_{L,n}[f](x)$ follows from the exactness of polynomials $P_{L,n}[f](x)$. □

Remark 7.3. The polynomial $\overline{P}_{L,n}[f](x)$ satisfies the interpolation conditions

$$\overline{P}_{L,n}[f](z) = f(z), \quad L(\overline{P}^{(i)}_{L,n}[f]) = L(f^{(i)}), \quad i = 1,\ldots,n. \tag{7.13}$$

Remark 7.4. Let $L(f) = f(x_0)$, $x_0 \in I$. The Appell interpolant polynomial associated to L is the known Taylor polynomial. Whereas for an arbitrary functional with $L(1) \neq 0$, relation (7.1) can be considered as a generalized Taylor polynomial. Other generalizations of Appell interpolation has been given in [63, 64]; generalizations of Sheffer polynomials [179] will be considered in the next chapter.

7.2 Operators approximation theory

The Appell polynomial sequence has been used in approximation theory with positive operators.

In fact, Jakimovski and Leviatan [112] consider the class of operators

$$P_n[f](x) := \frac{\exp(-nx)}{g(1)} \sum_{k=0}^{\infty} p_k(nx) f\left(\frac{k}{n}\right), \quad n \in \mathbb{N}, \quad n > 0, \tag{7.14}$$

for each function $f(x)$ defined in $[0,+\infty)$, where $\{p_k\}_k$ is the A. p. s. with the generating function $g(t)\exp(xt)$ and $g(1) \neq 0$.

Those operators generalize Szazs operators [190]:

$$P_n[f,z](x) := \exp(-nx)\sum_{k=0}^{\infty}\frac{(nx)^k}{k!}f\left(\frac{k}{n}\right),\quad n\in\mathbb{N},\quad n>0.$$

When $p_k(x) \geq 0$, $\forall x \in [0,+\infty)$, and the function $g(t)$ is holomorphic in the disc $|z| < R$, $R > 1$, the authors prove the following theorems.

Theorem 7.4 ([112]). *Suppose that $|f(t)| \leq \exp(At)$ for $t \geq 0$ for some finite A, that is, it is of exponential type. If $f(t)$ is continuous at $t = x_0$, then*

$$\lim_{n\to\infty} P_n[f](x_0) = f(x_0).$$

Theorem 7.5 ([112]). *Suppose that $f(t)$ is continuous in the infinite interval $[0,\infty)$ and that $\lim_{t\to\infty} f(t)$ exists. Then*

$$\lim_{n\to\infty} P_n[f](x) = f(x)$$

uniformly in $[0,\infty)$.

Theorem 7.6 ([112]). *Suppose that $f^{(r)}(t)$ exists for $0 \leq t < \infty$ and that*

$$|f(t)| \leq \exp(At) \quad \text{for} \quad t \geq 0$$

and some finite A. Then at each point $x = x_0$ where $f^{(r)}(x)$ is continuous,

$$\lim_{n\to\infty}\frac{d^r}{dx^r}P_n[f](x) = f^{(r)}(x_0).$$

Theorem 7.7 ([112]). *Suppose that $|f(t)| \leq \exp(At)$ for $t \geq 0$ and finite A. If $f'(x_0)$ exists for some $0 < x_0 < \infty$, then*

$$\lim_{n\to\infty}\frac{d}{dx}P_n[f](x) = f'(x_0).$$

Abel and Ivan [2] give an asymptotic expansion of the Jamikovski–Leviatan operators (7.14) and their derivatives:

$$P_n^{(l)}[f](x) \sim f(x)^{(l)} + \sum_{k=0}^{\infty} c_k^{(l)}(f,x)n^{-k} \quad (n\to\infty),\quad l = 0,1,\ldots, \tag{7.15}$$

where $c_k^{(l)}(f,x)$ will be defined later. If in addition to the hypotheses of Theorems 7.4–7.7, $f \in C^l(0,+\infty)$, $l \geq 0$, then it follows for $q > 0$ that

$$P_n^{(l)}[f](x) = f^{(l)}(x) + \sum_{k=1}^{q} c_k^{[l]}(f,x)n^{-k} + o(n^{-q}) \quad (n\to\infty), \tag{7.16}$$

where the coefficients $c_k^{[l]}(f,x)$ are given by

$$c_k^{[l]}(f,x) = \left(\frac{d}{dx}\right)^l c_k(f,x)$$

and

$$c_k(f,x) = \sum_{s=0}^{k} a(k,s)\frac{x^s}{s!} f^{(k+s)}(x)$$

with

$$a(k,s) = \frac{1}{k!}\sum_{m=0}^{k}\binom{k}{m}\frac{g^{(m)}(1)}{g(1)}\sum(-1)^{s-r}\frac{\binom{s}{r}}{\binom{r+k}{r+m}}S(r+k,r+m),$$

where $S(r,k)$ denote the Stirling numbers of the second kind.

We note that from (7.15), (7.16), the order of convergence of $P_n^{(r)}$ to $f^{(r)}$ is $\frac{1}{n}$. To improve the order of convergence of (7.16), we can apply the extrapolation procedure described in [45, 46, 140]. In fact, under the hypotheses of Theorems 7.4–7.7, we have the following:

Theorem 7.8 ([45, 46]). *Let $\{n_0 < n_1 < \cdots\}$ be an increasing sequence of positive integers, let $h_i = n_i^{-1}$, and let the sequence of operators be defined by*

$$T_0^{(i)} := T_0^{(i)}(f)(x) = P_{n_i}[f](x), \tag{7.17}$$

$$T_q^{(i)} := T_q^{(i)}(f)(x) = \frac{h_{i+q}T_{q-1}^{(i)} - h_i T_{q-1}^{(i+1)}}{h_{i+q} - h_i}, \qquad q = 1,\dots,h-1.$$

Then

$$\lim_{h_i\to 0} T_q^{(i)} = f(x). \tag{7.18}$$

More precisely, the following representations of $T_q^{(i)}$ hold:

$$T_q^{(i)}(f)(x) = f(x) + h_i h_{i+1} h_q(-1)^q(c_{q+1}(f,x) + o(h_i)), \tag{7.19}$$

$$T_q^{(i)}(f)(x) = \sum_{j=0}^{q} l_j(0)P_{n_i}[f](x), \quad l_j(h) = \prod_{\substack{i\neq j\\ i=0}}^{q}\frac{h_i - h}{h_i - h_j}. \tag{7.20}$$

In the following chapters, we will consider some examples of operators belonging to class (7.14), which satisfy the hypotheses of Theorems 7.4–7.8, but a more complete study has been presented in [55].

8 Examples

Abstract: In this chapter, we consider some important A. p. s. In particular, with previous techniques, we get Bernoulli, Euler, and Hermite polynomials. Moreover, we consider a new example, called the Appell–Laguerre polynomials.

8.1 Bernoulli polynomials

The known Bernoulli polynomial sequence is associated with the name of J. Bernoulli [24]. These polynomials play an important role in various expansion and approximation formulas, which are useful both in analytic theory of numbers and in classical and numerical analysis. Bernoulli polynomials can be defined in different ways depending on the applications. For a short review of approaches see [44] and references therein. Recently, two new closed forms for Bernoulli polynomials have been introduced in [159]. Relations with other polynomials have also been recently considered (see [86] and the references therein). Now we will follow the above theory, tracing a brief profile.

Let $(\widehat{a}_i)_i$ be the numerical sequence defined by

$$\widehat{a}_i = \frac{1}{i+1}, \quad i = 0, 1, \ldots, n, \ldots. \tag{8.1}$$

We get the matrix $\widehat{A} := (\widehat{a}_{i,j})$ with

$$\begin{cases} \widehat{a}_{i,j} = \binom{i}{j}\widehat{a}_{i-j} = \binom{i}{j}\frac{1}{i-j+1} = \frac{1}{i+1}\binom{i+1}{j}, & i \geq j, \\ \widehat{a}_{i,j} = 0, & i < j. \end{cases} \tag{8.2}$$

Then we set $A := (a_{i,j})$ with

$$a_{i,j} = \begin{cases} \binom{i}{j}a_{i-j}, & i \geq j, \\ 0, & i < j, \end{cases} \tag{8.3}$$

where $(a_i)_i$ satisfy (2.10). So we get

$$\frac{1}{i+1}\sum_{k=0}^{i}\binom{i+1}{k}a_k = \delta_{i,0}, \quad i = 0, 1, \ldots. \tag{8.4}$$

From (8.3) and (8.4), we have

$$\begin{cases} a_{i,i} = 1, & i = 0, 1, \ldots, \\ a_{i,j} = -\frac{1}{i-j+1}\binom{i}{j}\sum_{k=0}^{i-j-1}\binom{i-j+1}{k}a_k, & j < i, \\ a_{i,j} = 0, & j > i. \end{cases} \tag{8.5}$$

https://doi.org/10.1515/9783110652925-008

The matrices A and $\widehat{A}$ are Appell matrices, and from them, we define the conjugate A. p. s.:

$$\begin{aligned} a_0(x) &= a_{0,0}, & \widehat{a}_0(x) &= \widehat{a}_{0,0}, \\ a_1(x) &= a_{1,0} + a_{1,1}x, & \widehat{a}_1(x) &= \widehat{a}_{1,0} + \widehat{a}_{1,1}x, \\ &\cdots\cdots\cdots & &\cdots\cdots\cdots \\ a_n(x) &= a_{n,0} + a_{n,1}x + \cdots + a_{n,n}x^n, & \widehat{a}_n(x) &= \widehat{a}_{n,0} + \widehat{a}_{n,1}x + \cdots + \widehat{a}_{n,n}x^n, \\ &\cdots\cdots\cdots & &\cdots\cdots\cdots \end{aligned}$$

In particular, from (8.4), $\{a_n\}_n$ is known as the *Bernoulli polynomial sequence*, and it is indicated by $\{B_n\}_n$.[1] Moreover, to the knowledge of the author, the sequence $\{\widehat{a}_n\}_n$ is not noted in the literature. We call them the *Bernoulli conjugate sequence*, and we indicate them by $\{\tilde{B}_n\}_n$.

The first six polynomials are listed in Table 8.1.

Then in Figures 8.1 and 8.2, we compare the graphics of Bernoulli polynomials and of Bernoulli conjugate polynomials.

Table 8.1: Bernoulli and Bernoulli conjugate polynomials.

Bernoulli polynomials	**Bernoulli conjugate polynomials**
$B_0(x) = 1$	$\tilde{B}_0(x) = 1$
$B_1(x) = x - \frac{1}{2}$	$\tilde{B}_1(x) = x + \frac{1}{2}$
$B_2(x) = x^2 - x + \frac{1}{6}$	$\tilde{B}_2(x) = x^2 + x + \frac{1}{3}$
$B_3(x) = x^3 - \frac{3}{2}x^2 + \frac{1}{2}x$	$\tilde{B}_3(x) = x^3 + \frac{3}{2}x^2 + x + \frac{1}{4}$
$B_4(x) = x^4 - 2x^3 + x^2 - \frac{1}{30}$	$\tilde{B}_4(x) = x^4 + 2x^3 + 2x^2 + x + \frac{1}{5}$
$B_5(x) = x^5 - \frac{5}{2}x^4 + \frac{5}{3}x^3 - \frac{1}{6}x$	$\tilde{B}_5 = x^5 + \frac{5}{2}x^4 + \frac{10}{3}x^3 + \frac{5}{2}x^2 + x + \frac{1}{6}$

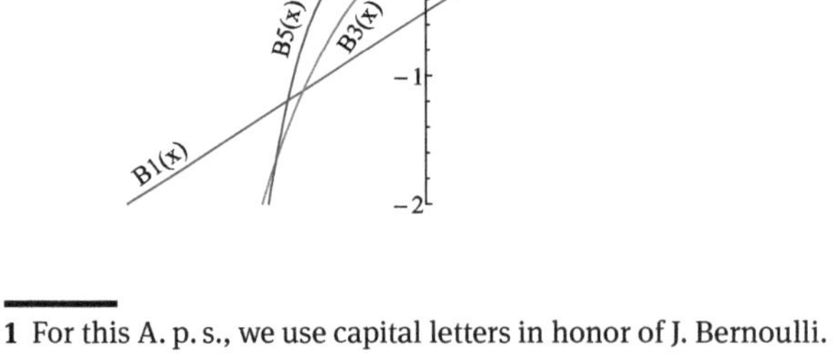

Figure 8.1: Bernoulli polynomials.

1 For this A. p. s., we use capital letters in honor of J. Bernoulli.

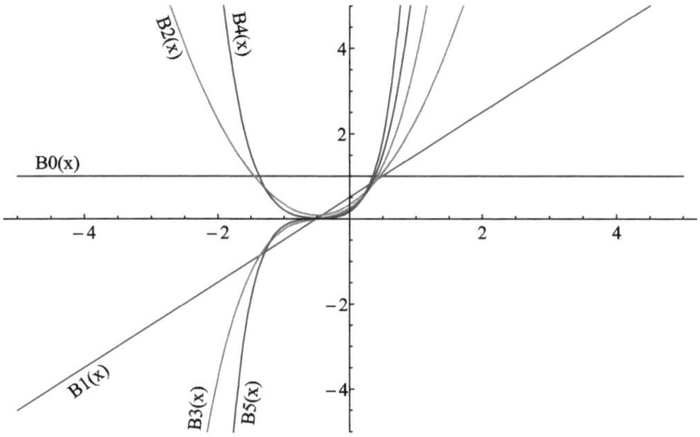

Figure 8.2: Bernoulli conjugate polynomials.

Table 8.2: Bernoulli and conjugate Bernoulli numbers.

Bernoulli numbers	Bernoulli conjugate numbers
$B_0 = 1$	$\tilde{B}_0 = 1$
$B_1 = -\frac{1}{2}$	$\tilde{B}_1 = \frac{1}{2}$
$B_2 = \frac{1}{6}$	$\tilde{B}_2 = \frac{1}{3}$
$B_3 = 0$	$\tilde{B}_3 = \frac{1}{4}$
$B_4 = -\frac{1}{30}$	$\tilde{B}_4 = \frac{1}{5}$
$B_5 = 0$	$\tilde{B}_5 = \frac{1}{6}$

We note that

$$B_n(x) = \sum_{k=0}^{n} \binom{n}{k} a_{n-k} x^k, \quad \tilde{B}_n(x) = \sum_{k=0}^{n} \binom{n}{k} \hat{a}_{n-k} x^k, \tag{8.6}$$

from which the coefficients are independent of the degree, and the result is

$$B_n := B_n(0) = a_n, \quad \tilde{B}_n := \tilde{B}_n(0) = \hat{a}_n. \tag{8.7}$$

Therefore, from (8.6) and (8.7), we have

$$B_n(x) = \sum_{k=0}^{n} \binom{n}{k} B_{n-k} x^k, \quad \tilde{B}_n(x) = \sum_{k=0}^{n} \binom{n}{k} \tilde{B}_{n-k} x^k. \tag{8.8}$$

The numbers B_n are known as *Bernoulli numbers* [34, 44], and we call $\tilde{B}_n$ the *conjugate Bernoulli numbers*. The first six Bernoulli and conjugate Bernoulli numbers are listed in Table 8.2.

The Bernoulli numbers can be calculated by recursive recurrence (8.4), but also by a closed formula [159], and they are applied not only in mathematics, [15, 88, 94].

The conjugate Bernoulli numbers are explicitly defined by

$$\tilde{B}_i = \frac{1}{i+1}, \quad i = 0, 1, \ldots,$$

and are connected to harmonic numbers [16]:

$$H_n = 1 + \frac{1}{2} + \cdots + \frac{1}{n} = \sum_{i=0}^{n-1} \tilde{B}_i.$$

For umbral composition, we have

$$B_n(\tilde{B}_n(x)) = \tilde{B}_n(B_n(x)) = x^n,$$

hence resulting in

$$\sum_{k=0}^{n} \binom{n}{k} B_{n-k}\tilde{B}_k(x) = \sum_{k=0}^{n} \binom{n}{k} \tilde{B}_{n-k}B_k(x) = x^n. \tag{8.9}$$

For the conjugate Bernoulli polynomial sequence $\{\tilde{B}_n\}_n$, we have

$$\tilde{B}_n(x) = \sum_{k=0}^{n} \binom{n}{k} \hat{a}_{n-k}x^k = \sum_{k=0}^{n} \binom{n}{k} \frac{1}{n-k+1} x^k = \frac{1}{n+1} \sum_{k=0}^{n} \binom{n+1}{k} x^k,$$

and hence

$$\tilde{B}_n(x) = \frac{1}{n+1}\left[(x+1)^{n+1} - x^{n+1}\right]. \tag{8.10}$$

From (8.10), we also have

$$\tilde{B}_n(x) = \frac{1}{n+1}\Delta x^{n+1} = \int_{x}^{x+1} t^n dt. \tag{8.11}$$

From Theorem 6.1, we get

$$B_n'(x) = nB_{n-1}(x), \quad \tilde{B}_n'(x) = n\tilde{B}_{n-1}(x), \quad n = 1, 2, \ldots, \tag{8.12}$$

and, consequently, the operational matrix of derivatives is

$$\begin{bmatrix} 0 & 0 & \cdots & 0 & 0 \\ 1 & 0 & \cdots & 0 & 0 \\ 0 & 2 & \ddots & 0 & 0 \\ \vdots & & \ddots & & \vdots \\ 0 & \cdots & & n & 0 \end{bmatrix}. \tag{8.13}$$

Integrating (8.12), we have

$$\int_0^x B_n(t)dt = \frac{1}{n+1}(B_{n+1}(x) - B_{n+1}), \quad \int_0^x \tilde{B}_n(t)dt = \frac{1}{n+1}\left(\tilde{B}_{n+1}(x) - \frac{1}{n+2}\right) \tag{8.14}$$

and

$$\int_0^1 B_n(t)dt = 0, \quad n > 1, \tag{8.15}$$

$$\int_0^1 \tilde{B}_n(t)dt = \frac{1}{(n+1)(n+2)}(2^{n+2} - 2). \tag{8.16}$$

The result is also

$$\int_0^1 B_n(t)B_m(t)dt = (-1)^{n-1}\frac{n!m!}{(n+m)!}B_{n+m}, \quad m, n \geq 1. \tag{8.17}$$

For the Bernoulli polynomials, the operational matrix [162] of integration (6.75) is

$$\mathcal{I} = \overline{U}\widehat{A}_n,$$

where

$$\overline{U} = [u_1, \ldots, u_n, W^T A]^T,$$

given that

$$u_i = \left[0, \frac{1}{i}\binom{i}{i-1}B_{i-1}, \ldots, \frac{1}{i}\binom{i}{0}B_0, \underbrace{0, \ldots, 0}_{n-i}\right]$$

and $W = [\mathcal{X}_0, \ldots, \mathcal{X}_n]$ with

$$\frac{B_{n+1}(x) - B_{n+1}}{n+1} \simeq \sum_{i=0}^{n} \mathcal{X}_i \widehat{A}_i(x).$$

For the conjugate Bernoulli polynomials, we have that

$$\mathcal{I} = \overline{U}A,$$

where

$$\overline{U} = [u_1, \ldots, u_n, W^T B]^T,$$

given that

$$u_i = \left[0, \frac{1}{i}\binom{i}{i-1}\frac{1}{i}, \frac{1}{i}\binom{i}{i-2}\frac{1}{i-1}, \dots, \frac{1}{i}\binom{i}{0}, \underbrace{0, \dots, 0}_{n-i}\right],$$

and $W = [\mathcal{X}_0, \dots, \mathcal{X}_n]$ with

$$\frac{(n+2)\tilde{B}_{n+1}(x) - 1}{(n+1)(n+2)} \simeq \sum_{i=0}^{n} \mathcal{X}_i \tilde{B}_i(x).$$

The Bernoulli operational matrices have recently been used in the numerical solution of differential and integral equations [1, 26, 109, 171, 195, 204, 216].

From the sequences (a_i) and $(\hat{a}_i)$, we get the formal power series

$$a(t) = \sum_{n=0}^{\infty} a_n \frac{t^n}{n!}, \quad \hat{a}(t) = \sum_{n=0}^{\infty} \hat{a}_n \frac{t^n}{n!}, \tag{8.18}$$

and we have

$$\hat{a}(t) = \sum_{n=0}^{\infty} \frac{1}{n+1} \frac{t^n}{n!} = \frac{\exp(t) - 1}{t}, \tag{8.19}$$

whereas for $a(t)$, given that $a(t)\hat{a}(t) = 1$, we have

$$a(t) = \frac{t}{\exp(t) - 1}. \tag{8.20}$$

Consequently,

$$G(x,t) = \frac{t}{\exp(t) - 1} \exp(xt), \quad \tilde{G}(x,t) = \frac{\exp(t) - 1}{t} \exp(xt) \tag{8.21}$$

are the generating functions of $\{B_n\}$ and $\{\tilde{B}_n\}$. In fact, we have

$$\frac{t}{\exp(t) - 1} \exp(xt) = \sum_{n=0}^{\infty} B_n(x) \frac{t^n}{n!}, \quad \frac{\exp(t) - 1}{t} \exp(xt) = \sum_{n=0}^{\infty} \tilde{B}_n(x) \frac{t^n}{n!}. \tag{8.22}$$

Directly from (6.26), we have the recurrence relations

$$B_n(x) = x^n - \sum_{k=0}^{n-1} \binom{n}{k} \frac{1}{n-k+1} B_k(x), \tag{8.23}$$

$$\tilde{B}_n(x) = x^n - \sum_{k=0}^{n-1} \binom{n}{k} B_{n-k} \tilde{B}_k(x). \tag{8.24}$$

For $n > 0$ and $x = 0$, the (8.23) gives the recurrence relation for the calculation of the Bernoulli numbers:

$$B_n = -\sum_{k=0}^{n-1} \binom{n}{k} \frac{1}{n-k+1} B_k = -\frac{1}{n+1} \sum_{k=0}^{n-1} \binom{n+1}{k} B_k, \quad B_0 = 1. \tag{8.25}$$

From (8.23) and (8.24), we get the inverse relations

$$\begin{cases} x^n = \frac{1}{n+1}\sum_{k=0}^{n}\binom{n+1}{k}B_k(x), \\ x^n = \sum_{k=0}^{n}\binom{n}{k}B_{n-k}\tilde{B}_k(x), \end{cases} \tag{8.26}$$

Using the conjugate Bernoulli vectors, we can rewrite these relations as

$$\begin{cases} B(x) = AX, \\ \tilde{B}(x) = \widehat{A}X. \end{cases} \Longleftrightarrow \begin{cases} X = \widehat{A}B(x), \\ X = A\tilde{B}(x). \end{cases} \Longleftrightarrow \begin{cases} \tilde{B}(x) = \widehat{A}^2B(x), \\ B(x) = A^2\tilde{B}(x). \end{cases} \tag{8.27}$$

The recurrence formula (8.23) is equivalent to the following determinantal form:

$$B_0(x) = 1,$$
$$B_n(x) = (-1)^n \begin{vmatrix} 1 & x & x^2 & \cdots & x^n \\ 1 & \frac{1}{2} & \frac{1}{3} & \cdots & \frac{1}{n+1} \\ 0 & 1 & \binom{2}{1}\frac{1}{2} & & \binom{n}{1}\frac{1}{n} \\ \vdots & & \ddots & \ddots & \\ 0 & & & 1 & \binom{n}{n-1}\frac{1}{2} \end{vmatrix}, \tag{8.28}$$

and, for $x = 0$,

$$B_n = (-1)^n \begin{vmatrix} 1 & \frac{1}{2} & \frac{1}{3} & \cdots & \frac{1}{n+1} \\ 0 & 1 & \binom{2}{1}\frac{1}{2} & & \binom{n}{1}\frac{1}{n} \\ \vdots & & \ddots & \ddots & \\ 0 & & & 1 & \binom{n}{n-1}\frac{1}{2} \end{vmatrix}. \tag{8.29}$$

For the conjugate Bernoulli polynomials, from relation (8.24), we have:

$$\tilde{B}_0(x) = 1,$$
$$\tilde{B}_n(x) = \frac{1}{n+1}[(x+1)^{n+1} - x^{n+1}] = (-1)^n \begin{vmatrix} 1 & x & x^2 & \cdots & x^n \\ B_0 & B_1 & B_2 & \cdots & B_n \\ 0 & 1 & \binom{2}{1}B_1 & & \binom{n}{1}B_{n-1} \\ \vdots & & \ddots & \ddots & \\ 0 & & & 1 & \binom{n}{n-1}B_1 \end{vmatrix}. \tag{8.30}$$

For $x = 0$, we get the interesting identity

$$\frac{1}{n+1} = \tilde{B}_n(0) = \begin{vmatrix} B_1 & B_2 & \cdots & \cdots & B_{n-1} & B_n \\ 1 & \binom{2}{1}B_1 & \binom{3}{1}B_2 & \cdots & \binom{n-1}{1}B_{n-2} & \binom{n}{1}B_{n-1} \\ 0 & 1 & & & \binom{n-1}{3}B_{n-3} & \binom{n}{3}B_{n-2} \\ \vdots & 0 & 1 & & \cdots & \vdots \\ 0 & \cdots & 0 & & 1 & \binom{n}{n-1}B_1 \end{vmatrix}. \tag{8.31}$$

For the calculation of the Hessenberg determinant (8.31), using the expansion with respect to the first row, we get

$$\frac{1}{n+1} = (-1)^n \sum_{i=1}^{n} d_i B_i$$

with

$$d_n = (-1)^{n-1}, \quad d_i = -\sum_{j=1}^{n-i} \binom{n}{j} B_j d_{n-j+1}, \quad i = n-1, \dots, 1.$$

So for the calculation of the Bernoulli number, we get the new recurrent formula

$$B_n = -\frac{1}{n+1} + (-1)^n \sum_{i=1}^{n-1} d_i B_i. \tag{8.32}$$

From (6.36) for Bernoulli and conjugate Bernoulli numbers, we get

$$b_i = \sum_{k=0}^{i} \binom{i}{k} \frac{B_{k+1}}{i-k+1}. \tag{8.33}$$

Therefore, the second recurrence relation (6.34) for the Bernoulli polynomials is

$$B_{n+1}(x) = (x+b_0)B_n(x) + \sum_{k=0}^{n-1} \binom{n}{k} b_{n-k} B_k(x). \tag{8.34}$$

So for the Bernoulli polynomials, we have the second determinantal form

$$B_{n+1}(x) = \begin{vmatrix} x+b_0 & -1 & 0 & & \cdots & & 0 \\ b_1 & x+b_0 & -1 & & \cdots & & 0 \\ b_2 & \binom{2}{1} b_1 & x+b_0 & -1 & \cdots & & 0 \\ \vdots & & & \ddots & & & \vdots \\ b_{n-1} & \binom{n-1}{1} b_{n-2} & \binom{n-1}{2} b_{n-3} & & \cdots & x+b_0 & -1 \\ b_n & \binom{n}{1} b_{n-1} & \binom{n}{2} b_{n-2} & & \cdots & & x+b_0 \end{vmatrix}. \tag{8.35}$$

From (8.35), by applying Theorem 6.21, we have that the zeros of $B_n(x)$ lie inside the circle of center $b_0 = -\frac{1}{2}$ and radius r, where

$$r = \max_{0 \le i \le n} \sum_{\substack{j=0 \\ i \ne j}}^{i+1} \binom{i}{k} \left| \sum_{k=0}^{i-j} \binom{i-j}{k} \frac{B_{k+1}}{i-j+k+1} \right|. \tag{8.36}$$

Likewise, for the conjugate Bernoulli polynomials, given

$$\tilde{b}_i = \sum_{k=0}^{i} \binom{i}{k} \frac{B_{i-k}}{k+2}, \tag{8.37}$$

we get the second recurrence relation

$$\tilde{B}_{n+1}(x) = (x + \tilde{b}_0)\tilde{B}_n(x) + \sum_{k=0}^{n-1} \binom{n}{k} \tilde{b}_{n-k}\tilde{B}_k(x) \tag{8.38}$$

and the related determinantal form

$$\tilde{B}_{n+1}(x) = \begin{vmatrix} x + \tilde{b}_0 & -1 & 0 & & \cdots & & 0 \\ \tilde{b}_1 & x + \tilde{b}_0 & -1 & & \cdots & & 0 \\ \tilde{b}_2 & \binom{2}{1}\tilde{b}_1 & x + \tilde{b}_0 & -1 & \cdots & & 0 \\ \vdots & & & \ddots & & & \vdots \\ \tilde{b}_{n-1} & \binom{n-1}{1}\tilde{b}_{n-2} & \binom{n-1}{2}\tilde{b}_{n-3} & & \cdots & x + \tilde{b}_0 & -1 \\ \tilde{b}_n & \binom{n}{1}\tilde{b}_{n-1} & \binom{n}{2}\tilde{b}_{n-2} & & \cdots & & x + \tilde{b}_0 \end{vmatrix}. \tag{8.39}$$

From (8.39), by applying Theorem 6.21, we have that the zeros of $\tilde{B}_n(x)$ lie inside the circle of center $\tilde{b}_0 = -\frac{1}{2}$ and radius $\tilde{r}$, where

$$\tilde{r} = \max_{0 \le i \le n} \sum_{\substack{j=0 \\ i \ne j}}^{i+1} \binom{i}{k} \left| \sum_{k=0}^{i-j} \binom{i-j}{k} \frac{B_{i-j-k}}{k+2} \right|. \tag{8.40}$$

The conjugate Bernoulli polynomial give a basis of $\mathcal{P}_n$, then directly from Corollary 6.3, if

$$p_n(x) = \sum_{k=0}^{n} c_{n,k} x^k \in \mathcal{P}_n,$$

we get

$$p_n(x) = \sum_{k=0}^{n} \tilde{c}_{n,k} B_k(x) = \sum_{k=0}^{n} c^*_{n,k} \tilde{B}_k(x), \tag{8.41}$$

where

$$\tilde{c}_{n,k} = \frac{1}{k+1} \sum_{j=0}^{n-k} \binom{k+j}{k} c_{n,k+j}, \quad c^*_{n,k} = \sum_{j=0}^{n-k} \binom{k+j}{k} c_{n,k+j} B_k. \tag{8.42}$$

Moreover, if $\{y_n\}_n$ is the polynomial sequence

$$y_n(x) = \sum_{k=0}^{n} \binom{n}{k} \frac{1}{n-k+1} x^k,$$

it follows that

$$x^n = \sum_{k=0}^{n} \binom{n}{k} B_{n-k} y_k(x). \tag{8.43}$$

Likewise, from the numerical sequence $(y_n)_n$ with

$$y_n = \sum_{k=0}^{n} \binom{n}{k} \frac{1}{n-k+1} x_k,$$

it follows that

$$x_n = \sum_{k=0}^{n} \binom{n}{k} B_{n-k} y_k. \tag{8.44}$$

We note explicitly that the matrices with entries

$$x_{n,k} = \binom{n}{k} B_{n-k} \quad \text{and} \quad y_{n,k} = \frac{1}{n+1} \binom{n+1}{k}$$

are inverse of each other.

From (6.52), we have the following identities:

$$B_n(x+y) = \sum_{i=0}^{n} \binom{n}{i} B_i(x) y^{n-i} = \sum_{i=0}^{n} \binom{n}{i} B_{n-i}(y) x^i, \tag{8.45}$$

$$\tilde{B}_n(x+y) = \sum_{i=0}^{n} \binom{n}{i} \tilde{B}_i(x) y^{n-i}. \tag{8.46}$$

For $y = -x$, we get

$$\sum_{i=0}^{n} \binom{n}{i} B_i(x)(-1)^{n-i} x^{n-i} = \begin{cases} 0 & \text{for } n \text{ odd, } n > 1, \\ B_n & \text{for } n \text{ even.} \end{cases} \tag{8.47}$$

Now, by setting $y = 1$ in (8.45), we have

$$B_n(x+1) = \sum_{i=0}^{n} \binom{n}{i} B_i(x),$$

that is,

$$B_n(x+1) = B_n(x) + \sum_{i=0}^{n-1} \binom{n}{i} B_i(x) = B_n(x) + n \sum_{i=0}^{n-1} \binom{n-1}{i} \frac{1}{n-i} B_i(x) = B_n(x) + nx^{n-1},$$

and so

$$\Delta B_n(x) \equiv B_n(x+1) - B_n(x) = nx^{n-1}. \tag{8.48}$$

From (8.48), for $n > 1$ and $x = 0$, we have

$$B_n(1) - B_n(0) = 0,$$

that is,

$$B_n(1) = B_n, \quad n > 1. \tag{8.49}$$

Some authors, for example, [113, p. 231], assumed (8.48) with the first of (8.12) as the definition of Bernoulli polynomials.

Moreover, from (8.48), we get

$$B_n(1-x) = (-1)^n B_n(x) \tag{8.50}$$

and, for odd n and $x = 0$,

$$B_n(1) = -B_n. \tag{8.51}$$

Therefore, from (8.49) and (8.51), for odd n and $n > 1$, we get

$$B_n = 0. \tag{8.52}$$

A proof of (8.50) also follows from the first determinantal form. Moreover, a direct proof of (8.52) by determinantal form (8.29) is not known in the literature.

From (8.50), we also get

$$B_{2n}(1-x) = B_{2n}(x), \tag{8.53}$$

and putting $x = \frac{1}{2} + z$, we obtain

$$B_{2n}\left(\frac{1}{2} + z\right) = B_{2n}\left(\frac{1}{2} - z\right). \tag{8.54}$$

Using (8.54), we can determine the values of Bernoulli polynomials of even degree in the interval $(\frac{1}{2}, 1)$, if those values are known in the interval $(0, \frac{1}{2})$. So the curve $y = B_{2n}(x)$ is symmetric with respect to the line $x = \frac{1}{2}$ in the interval $(0, 1)$.

The differentiation of (8.54) gives

$$B_{2n-1}\left(\frac{1}{2} + z\right) = -B_{2n-1}\left(\frac{1}{2} - z\right), \tag{8.55}$$

and we conclude that Bernoulli polynomials of odd degree are symmetric with respect to the point $(\frac{1}{2}, 0)$.

Putting $z = 0$ in (8.55), we have

$$B_{2n-1}\left(\frac{1}{2}\right) = -B_{2n-1}\left(\frac{1}{2}\right), \tag{8.56}$$

hence $B_{2n-1}(\frac{1}{2}) = 0$, $\forall n > 1$. Therefore, the polynomials $B_{2n-1}(x)$ with $n > 1$ have three roots in $[0, 1]$: $x = 0$, $x = \frac{1}{2}$, $x = 1$. In fact, in [113, p. 239], it is proved that the polynomials of odd degree cannot have more than three roots in the interval $0 \le x \le 1$.

Now some particular values of Bernoulli polynomials are given:

$$\begin{cases} B_n(0) = B_n(1) = B_n, & n > 1, \\ B_n(\frac{1}{2}) = (2^{1-n} - 1)B_n, & n \geq 2, \\ B_n(\frac{1}{6}) = \frac{1}{2}(1 - 2^{-2n+1})(1 - 3^{-2n+1})B_{2n}. \end{cases} \tag{8.57}$$

For $k \geq 1$, the following relation holds:

$$(-1)^k B_{2k}(x) > 0 \quad \forall x \in \left(0, \frac{1}{2}\right).$$

Bounds for Bernoulli polynomials and numbers are well known in the literature (see [15, 94, 123, 124] and references therein).

Some of them are:

$$|B_{2n}(x)| < |B_{2n}|, \tag{8.58}$$

$$4\sqrt{\pi n}\left(\frac{n}{\pi}\right)^{2n} < |B_{2n}| < 5\sqrt{\pi n}\left(\frac{n}{\pi}\right)^{2n}. \tag{8.59}$$

For the zeros of Bernoulli polynomials, there is a wide literature (see [84, 123] and reference in therein). The upper bound [85]

$$\left|z - \frac{1}{2}\right| < \sqrt{\frac{n(n-1)}{2n}} \tag{8.60}$$

holds.

For Bernoulli conjugate sequence $\{\tilde{B}_n(x)\}_n$, we have the following results:

1. From (8.10), it follows that

$$\tilde{B}_{2k}(x) > 0 \quad \forall x \in \mathbb{R}, \quad \tilde{B}_{2k-1}(x) > 0 \quad \forall x > 0, \tag{8.61}$$

   and $x = -\frac{1}{2}$ is the unique real zero of $\tilde{B}_{2k+1}(x)$, that is,

$$\tilde{B}_{2k+1}(x) = 0 \quad \text{only for} \quad x = -\frac{1}{2}, \qquad x \in \mathbb{R}. \tag{8.62}$$

2. $\tilde{B}_{2k}(x)$ has a unique minimum point for $x = -\frac{1}{2}$. In fact,

$$\tilde{B}'_{2k}(x) = 2k\tilde{B}_{2k-1}(x), \quad \tilde{B}''_{2k}(x) = 2k(2k-1)\tilde{B}_{2k-2}(x),$$

   and the result follows from (8.61) and (8.62).

3. The result is

$$\tilde{B}_{2k}\left(-\frac{1}{2}\right) = \frac{1}{2k+1}\left\{\left(\frac{1}{2}\right)^{2k+1} + \left(\frac{1}{2}\right)^{2k+1}\right\} = \frac{1}{(2k+1)2^{2k}},$$

   and, for item 2, we have

$$\tilde{B}_{2k}(x) \geq \frac{1}{(2k+1)2^{2k}}. \tag{8.63}$$

4. From item 2, it follows that $\tilde{B}_{2k}(x)$ is symmetric with respect to the straight line $x = -\frac{1}{2}$, that is,

$$\tilde{B}_{2k}\left(x - \frac{1}{2}\right) = \tilde{B}_{2k}\left(-x - \frac{1}{2}\right). \tag{8.64}$$

5. We get the following forward difference

$$\begin{aligned}\Delta\tilde{B}_n(x) &= \tilde{B}_n(x+1) - \tilde{B}_n(x)\\ &= \frac{1}{n+1}\{(x+2)^{n+1} - 2(x+1)^{n+1} + x^{n+1}\}\\ &= \frac{1}{n+1}\Delta^2 x^{n+1}.\end{aligned} \tag{8.65}$$

Moreover, we get that

$$\begin{aligned}\tilde{B}_n(x+1) &= \sum_{i=0}^{n}\binom{n}{i}\tilde{B}_i(x) = \tilde{B}_n(x) + \sum_{i=0}^{n-1}\binom{n}{i}\tilde{B}_i(x)\\ &= \tilde{B}_n(x) + n\sum_{i=0}^{n-1}\binom{n-1}{i}\frac{1}{n-i}\tilde{B}_i(x),\end{aligned}$$

from which

$$\Delta\tilde{B}_n(x) = n\sum_{i=0}^{n-1}\binom{n-1}{i}\frac{1}{n-i}\tilde{B}_i(x).$$

We observe that

$$\hat{a}_n = \frac{1}{n+1} = \int_0^1 x^n dx, \tag{8.66}$$

from which $\{B_n\}_n$ is the A. p. s. relative to the linear functional

$$L(x^i) = \int_0^1 x^i dx, \quad i = 0, 1, \ldots. \tag{8.67}$$

From Theorem 6.28, for any $P_n(x) \in \mathcal{P}_n$, we have

$$P_n(x) = \int_0^1 P_n(x)dx + \sum_{k=1}^{n}\frac{(P_n^{(k-1)}(1) - P_n^{(k-1)}(0))}{k!}B_k(x). \tag{8.68}$$

From (8.48), we get

$$B_n(x) = n\Delta^{-1}x^{n-1} + k \tag{8.69}$$

with a real constant k, leading to the assumption that, for $n > 1$,

$$a_n = B_n = B_n(0) = (n\Delta^{-1}x^{n-1})_{x=0}. \qquad (8.70)$$

In this way, the conjugate Bernoulli sequence $\{\tilde{B}_n\}$ is the A. p. s. related to the linear functional

$$\tilde{L}(x^i) = i(\Delta^{-1}x^i)_{x=0}, \quad i = 1, 2, \ldots. \qquad (8.71)$$

From Theorem 6.28, for any $P_n(x) \in \mathcal{P}_n$, we have

$$P_n(x) = \sum_{k=0}^{n} \frac{\tilde{L}(P_n^{(k)}(x))}{k!}\tilde{B}_n(x). \qquad (8.72)$$

Let $\mathbf{X}$ be the linear space of real and continuous functions defined in the interval $[0, 1]$ and with continuous derivatives of all necessary orders. Then we consider the polynomials

$$P_{L,n}[f](x) = \sum_{i=0}^{n} \frac{L(f^{(i)})}{i!}B_n(x), \qquad (8.73)$$

$$\tilde{P}_{\tilde{L},n}[f](x) = \sum_{i=0}^{n} \frac{\tilde{L}(f^{(i)})}{i!}\tilde{B}_n(x), \qquad (8.74)$$

and using Theorem 7.1, we have

$$L(P_{L,n}[f]^{(i)}(x)) = L(f^{(i)}), \quad i = 0, \ldots, n, \qquad (8.75)$$

$$\tilde{L}(\tilde{P}_{\tilde{L},n}[f]^{(i)}(x)) = \tilde{L}(f^{(i)}), \quad i = 0, \ldots, n. \qquad (8.76)$$

Moreover, from Theorems 7.1, 7.3, 7.4, we obtain the following:

Theorem 8.1 ([43]). *If $f(x)$ is a real function of class C^ν, $\nu \geq 1$ in the interval $[0, 1]$, then for any $x \in [0, 1]$, the following relation holds:*

$$f(x) = f(0) + \sum_{k=0}^{\nu} \frac{(B_k(x) - B_k)}{k!}(f^{(k-1)}(1) - f^{(k-1)}(0)) + R_\nu[f](x) \qquad (8.77)$$

with

$$R_\nu[f](x) = \frac{1}{\nu!}\int_0^1 f^{(\nu)}(t)(B_\nu^*(x-t) + (-1)^{\nu+1}B_\nu(t))dt \qquad (8.78)$$

given $B_\nu^*(x)$, the periodic Bernoulli function defined by

$$\begin{cases} B_\nu^*(x) = B_\nu(x), & 0 \leq x < 1, \\ B_\nu^*(x+1) = B_\nu^*(x) & \text{otherwise.} \end{cases}$$

This theorem has been applied in different contexts, in particular,
1. for interpolation of real functions [67];
2. for numerical solutions of nonlinear equations [59],
3. for numerical solutions of BVPs by collocation methods [69, 71].

Moreover, it has been generalized to bivariate functions in triangular and rectangular domain [53].

In [28], the authors consider the expansion of a function of exponential type in the Bernoulli polynomials, that is, they consider the series

$$f(x) \sim \int_0^1 f(x)dx + \sum_{n=1}^{\infty} [f^{(n-1)}(1) - f^{(n-1)}(0)] \frac{B_n(x)}{n!} \tag{8.79}$$

and proved the uniform convergence of it, but in [68], there is a more elementary proof of this result.

Analogous theorems are true for the conjugate Bernoulli polynomials, but they seem to be more theoretical than practical due to the difficulty in calculation of the functional $\tilde{L}$.

On the contrary, the polynomials $\{\tilde{B}_n\}_n$ are very useful in the approximation of real regular functions on the interval $[0, +\infty[$. In fact, denoting by $E(A)$ the class of the functions of exponential type, that is, with the property that $|f(t)| \leq e^{At}$ for each $t \geq 0$ and some finite number A, we have the following:

Theorem 8.2 ([55]). *Let $f \in C[0, \infty[\cap E(A)$, and let $P_n^{(l)}(f)(x)$ be the sequence of positive linear functionals*

$$P_n^{(l)}(f)(x) = \frac{\exp(-nx)}{\exp(1) - 1} \sum_{k=0}^{\infty} \tilde{B}_k^{(l)}(nx) f\left(\frac{k}{n}\right),$$

where $l \geq 0$ denotes the order of derivative. Then

$$\lim_{n\to\infty} P_n^{(l)}(f)(x) = f^{(l)}(x) \quad \forall x \in [0, +\infty[, \quad l = 0, 1, \ldots, n,$$

and the convergence is uniform in each compact $[0, k]$, $k > 0$.

Proof. It follows using Theorems 7.4–7.7, by considering the properties of $\tilde{B}_n(x)$. □

The analogous theorem for $B_k(x)$ is not true.

Moreover, using Theorem 7.8, we have:

Theorem 8.3 ([55]). *Let $\{n_0 < n_1 < \cdots\}$ be an increasing sequence of positive integers and $h_i = n_i^{-1}$. We define a sequence of operators of degree n_{i+q} as*

$$\begin{aligned} T_0^{(i)} &:= T_0^{(i)}(f)(x) = P_{n_i}^{(l)}(f)(x), \quad l = 0, 1 \ldots, \\ T_q^{(i)} &:= T_q^{(i)}(f)(x) = \frac{h_{i+q} T_{q-1}^{(i)} - h_i T_{q-1}^{(i+1)}}{h_{i+q} - h_i}, \quad q = 1, \ldots, h_{i-1}. \end{aligned} \tag{8.80}$$

Then

$$\lim_{h_i \longrightarrow 0} T_q^{(i)} = f^{(l)}(x).$$

More precisely, the following representation of $T_q^{(i)}$ *holds:*

$$T_q^{(i)}[f](x) = f(x) + h_i h_{i+1} \cdots h_q (-1)^q (C_{q+1}(f, x) + o(h_i)).$$

We note explicitly that the operator of Theorem 8.2 for $l = 0$ becomes

$$P_n[f](x) = \frac{\exp(-nx)}{(\exp(1) - 1)} \sum_{k=0}^{\infty} \frac{1}{k+1} [(nx+1)^{k+1} - (nx)^k] f\left(\frac{k}{n}\right). \tag{8.81}$$

The Bernoulli polynomials have been generalized in different ways, depending also on applications; for example, see [30, 67, 86, 88, 119, 136, 144, 181]. The classic generalization are the Bernoulli polynomials [166, p. 93] of order a with $a \neq 0$, defined by the generating function

$$\left(\frac{t}{\exp(t) - 1}\right)^a \exp(xt) = \sum_{n=0}^{\infty} B_n^{(a)}(x) \frac{t^n}{n!}.$$

In recent years, Bernoulli polynomials with weight [43] related to the linear functional

$$\widehat{a}_n^w = L(x^n) = \int_0^1 w(x) x^n dx$$

have been studied.

Now we recall some historical notes.

The term Bernoulli polynomials was used for the first time in 1851 by Raabe [161] in connection with the multiplication theorem

$$\frac{1}{m} \sum_{k=0}^{m-1} B_n\left(x + \frac{k}{m}\right) = m^{-n} B_n(mx).$$

J. Bernoulli introduced the polynomials $B_n(m)$ in 1690 (his work was published in 1713 [24]), jointly with the discovery of numbers B_n related to the calculation of the sum of the powers of the first natural numbers:

$$S_n(m) = \sum_{k=0}^{m-1} k^n.$$

He introduced the formula

$$S_n(m) = \sum_{k=0}^{n} \frac{B_k}{k!} \frac{n!}{(n+1-k)!} m^{n+1-k} = \sum_{k=0}^{n} \binom{n}{k} a_k \widehat{a}_{n-k} m^{n+1-k},$$

and he set

$$S_n(m) = \frac{1}{n+1}(B_{n+1}(m) - B_{n+1}(0)).$$

The successive approaches to Bernoulli polynomials have been made by L. Euler [89] and P. E. Appell [17]. In 1980, A. Hurwitz—in a personal communication to G. Polya—gave the Fourier series expansion for $B_n(x)$:

$$B_n(x) = \frac{n!}{(2\pi i)^n} \sum_{k=-\infty}^{+\infty} k^{-n} \exp(2\pi ikx), \quad 0 < x < 1,$$

and used the Fourier series approach to Bernoulli polynomials in lectures. In 1891, Lucas [131] derived the Bernoulli polynomials sequence using the umbral calculus:

$$B_n(x) = (B + x)^n.$$

He claimed that k as exponent of B^k in the power expansion of the right member can become the index of Bernoulli number B_k to obtain

$$(B + x)^n = \sum_{k=0}^{n} \binom{n}{k} B^k x^{n-k} = \sum_{k=0}^{n} \binom{n}{k} B_k x^{n-k}.$$

More recently, Lehmer [124] used the Raabe multiplication theorem in a new approach to Bernoulli polynomials and from this derived the other definitions. Finally, the approach of Costabile [43, 44, 62, 67] was inspired by an interpolation problem.

8.2 Euler polynomials

Euler polynomial sequence is a very well-known Appell sequence. Therefore, we sketch out a short profile.

Let $\hat{a}_i$, $i = 0, 1, \ldots$, be the numerical sequence defined by

$$\hat{a}_0 = 1, \quad \hat{a}_i = \frac{1}{2}, \quad i = 1, 2, \ldots. \tag{8.82}$$

Then we consider the infinite triangular matrix $\hat{A} = (\hat{a}_{i,j})$ with

$$\hat{a}_{i,j} = \binom{i}{j} \hat{a}_{i-j}, \quad i = 0, 1, \ldots, \quad j = 0, \ldots, i. \tag{8.83}$$

Now we consider the matrix A defined by

$$A = \hat{A}^{-1}, \tag{8.84}$$

and putting $A = (a_{i,j})$, we have

$$a_{i,j} = \binom{i}{j} a_{i-j}, \tag{8.85}$$

where the sequence $(a_i)_i$ satisfies

$$\sum_{k=0}^{i} \binom{i}{k} \hat{a}_{i-k} a_k = \delta_{i,0}, \quad i = 0, 1, \dots, \tag{8.86}$$

which is equivalent to

$$\sum_{k=0}^{i-1} \binom{i}{k} \frac{a_k}{2} + a_i = \begin{cases} 1, & i = 0, \\ 0, & i > 0. \end{cases} \Leftrightarrow \begin{cases} a_0 = 1, \\ a_i = -\frac{1}{2} \sum_{j=0}^{i-1} \binom{i}{j} a_j, & i > 0. \end{cases}$$

The matrix A is a nonsingular, infinite triangular, and it is an Appell matrix, like $\hat{A}$. Consequently, we may consider the conjugate A. p. s. $\{a_n\}_n$ and $\{\hat{a}_n\}_n$:

$$\begin{aligned} &a_0(x) = 1, && \hat{a}_0(x) = 1, \\ &a_1(x) = a_{1,0} + a_{1,1}x, && \hat{a}_1(x) = \hat{a}_{1,0} + \hat{a}_{1,1}x, \\ &\dots\dots\dots && \dots\dots\dots \\ &a_n(x) = a_{n,0} + a_{n,1}x + \cdots + a_{n,n}x^n, && \hat{a}_n(x) = \hat{a}_{n,0} + \hat{a}_{n,1}x + \cdots + \hat{a}_{n,n}x^n, \\ &\dots\dots\dots && \dots\dots\dots \end{aligned}$$

The sequence $\{a_n\}_n$ is known as the *Euler polynomial sequence* and is indicated as $\{E_n\}_n$ [62, 166]. The conjugate sequence $\{\hat{a}_n\}$ is unknown in literature, and we indicate it as $\{\tilde{E}_n\}_n$.

In Table 8.3, the first six Euler and Euler conjugate polynomials are listed.

Figures 8.3 and 8.4 show the graphics of these polynomials.

We observe that

$$\begin{cases} E_n(0) = a_{n,0} = a_n, \\ \tilde{E}_n(0) = \hat{a}_{n,0} = \hat{a}_n \begin{bmatrix} 1, & n = 0, \\ \frac{1}{2}, & n \geq 1, \end{bmatrix} \end{cases} \tag{8.87}$$

Table 8.3: Euler and Euler conjugate Polynomials.

Euler polynomials	**Euler conjugate polynomials**
$E_0(x) = 1$	$\tilde{E}_0(x) = 1$
$E_1(x) = x - \frac{1}{2}$	$\tilde{E}_1(x) = x + \frac{1}{2}$
$E_2(x) = x^2 - x$	$\tilde{E}_2(x) = x^2 + x + \frac{1}{2}$
$E_3(x) = x^3 - \frac{3}{2}x^2 + \frac{1}{4}$	$\tilde{E}_3(x) = x^3 + \frac{3}{2}x^2 + \frac{3}{2}x + \frac{1}{2}$
$E_4(x) = x^4 - 2x^3 + x$	$\tilde{E}_4(x) = x^4 + 2x^3 + 3x^2 + 2x + \frac{1}{2}$
$E_5(x) = x^5 - \frac{5}{2}x^4 + \frac{5}{2}x^2 - \frac{1}{2}$	$\tilde{E}_5 = x^5 + \frac{5}{2}x^4 + 5x^3 + 5x^2 + \frac{5}{2}x + \frac{1}{2}$

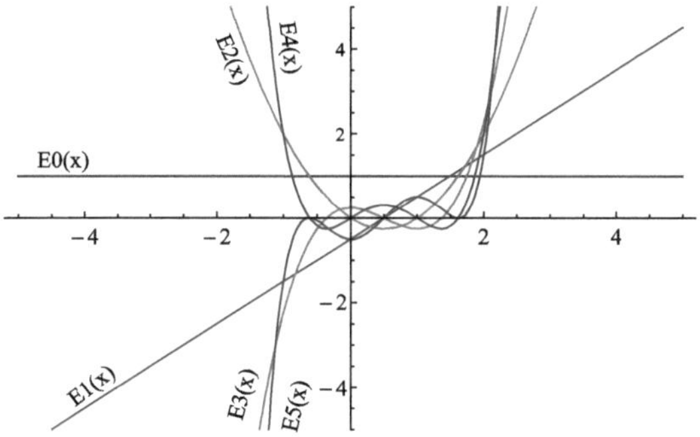

Figure 8.3: Euler polynomials.

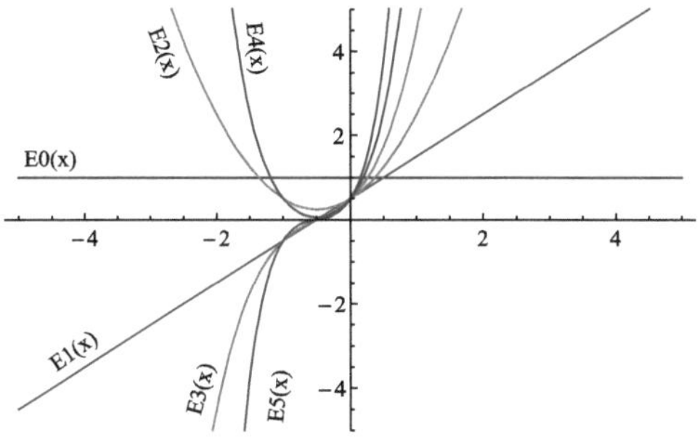

Figure 8.4: Euler conjugate polynomials.

and so we can write

$$E_n(x) = \sum_{k=0}^{n} \binom{n}{k} E_{n-k}(0)x^k, \quad \tilde{E}_n(x) = \frac{1}{2}[(x+1)^n + x^n]. \tag{8.88}$$

We have directly, by umbral composition, that

$$E_n(\tilde{E}_n(x)) = \tilde{E}_n(E_n(x)) = x^n.$$

From Theorem 6.1, we have

$$E_n'(x) = nE_{n-1}(x), \quad \tilde{E}_n'(x) = n\tilde{E}_{n-1}(x), \quad n = 1, 2, \ldots, \tag{8.89}$$

and by integration, we get

$$\int_0^x E_n(t)dt = \frac{1}{n+1}(E_{n+1}(x) - E_{n+1}(0)), \quad \int_0^x \tilde{E}_n(t)dt = \frac{1}{n+1}\left(\tilde{E}_{n+1}(x) - \frac{1}{2}\right). \tag{8.90}$$

For definite integral, we get

$$\int_0^1 E_n(t)dt = \frac{1}{n+1}(E_{n+1}(1) - E_{n+1}(0)), \quad \int_0^1 \tilde{E}_n(t)dt = \frac{2^n}{n+1}. \tag{8.91}$$

Consequently, the integration matrix is

$$\mathcal{I} = \overline{U}\hat{A}_n, \tag{8.92}$$

where

$$\overline{U} = [U_1, \dots, U_n, W^T A_n]^T,$$

given that

$$U_i = \left[0, \frac{1}{i}\binom{i}{i-1}E_{i-1}, \dots, \frac{1}{i}\binom{i}{0}E_0, \underbrace{0, \dots, 0}_{n-i}\right],$$

and $W = [\mathcal{X}_0, \dots, \mathcal{X}_n]$ in

$$\frac{E_{n+1}(x) - E_{n+1}(0)}{n+1} \simeq \sum_{i=0}^{n} \mathcal{X}_i E_i(x).$$

For the conjugate Euler polynomials, we have that

$$\mathcal{I} = \overline{U}A_n,$$

where

$$\overline{U} = [U_1, \dots, U_n, W^T \hat{A}_n]^T,$$

given that

$$U_i = \left[0, \frac{1}{2i}\binom{i}{i-1}, \frac{1}{2i}\binom{i}{i-2}, \dots, \frac{1}{i}\binom{i}{0}, \underbrace{0, \dots, 0}_{n-i}\right],$$

and $W = [\mathcal{X}_0, \dots, \mathcal{X}_n]$ in

$$\frac{\tilde{E}_{n+1}(x) - \frac{1}{2}}{n+1} \simeq \sum_{i=0}^{n} \mathcal{X}_i \tilde{E}_i(x).$$

If we define the power series

$$a(t) = \sum_{n=0}^{\infty} a_n \frac{t^n}{n!}, \quad \hat{a}(t) = \sum_{n=0}^{\infty} \hat{a}_n \frac{t^n}{n!},$$

we have

$$\hat{a}(t) = 1 + \frac{1}{2}\sum_{n=1}^{\infty} \frac{t^n}{n!} = 1 + \frac{1}{2}(\exp(t) - 1) = \frac{\exp(t) + 1}{2}. \tag{8.93}$$

Since $a(t)\hat{a}(t) = 1$, it follows that

$$a(t) = \frac{2}{\exp(t) + 1}. \tag{8.94}$$

Then the generating functions for Euler conjugate polynomials sequences are

$$\begin{aligned} G(x,t) &= \sum_{n=0}^{\infty} E_n(x)\frac{t^n}{n!} = \frac{2\exp(xt)}{\exp(t)+1}, \\ \tilde{G}(x,t) &= \sum_{n=0}^{\infty} \tilde{E}_n(x)\frac{t^n}{n!} = \frac{(\exp(t)+1)\exp(xt)}{2}. \end{aligned} \tag{8.95}$$

From (8.94), we have

$$a(t) = \sum_{n=0}^{\infty} a_n \frac{t^n}{n!} = 1 + \sum_{n=1}^{\infty} a_n \frac{t^n}{n!} = \frac{2}{\exp(t)+1}.$$

So

$$\sum_{n=1}^{\infty} a_n \frac{t^n}{n!} = \frac{2}{\exp(t)+1} - 1 = -\frac{\exp(t)-1}{\exp(t)+1} = -\tanh\frac{1}{2}t.$$

Recalling that

$$\tanh x = \sum_{n=1}^{\infty} 2^{2n}(2^{2n}-1)B_{2n}\frac{x^{2n-1}}{(2n)!}, \quad |x| < \frac{\pi}{2},$$

we have

$$\tanh\frac{x}{2} = \sum_{n=1}^{\infty} \frac{(2^{2n}-1)}{n}B_{2n}\frac{x^{2n-1}}{(2n-1)!}, \quad |x| < \frac{\pi}{2},$$

and comparing with the previous equalities, we have

$$a_{2n-1} = -\frac{2^{2n}-1}{n}B_{2n} \tag{8.96}$$

and

$$a_{2n} = 0. \tag{8.97}$$

Directly from (6.26), we have the recurrence relations

$$E_n(x) = x^n - \frac{1}{2}\sum_{k=0}^{n-1}\binom{n}{k}E_k(x), \tag{8.98}$$

$$\tilde{E}_n(x) = x^n - \sum_{k=0}^{n-1}\binom{n}{k}E_{n-k}(0)\tilde{E}_k(x). \tag{8.99}$$

Then from (8.98) for $n > 0$ and $x = 0$, it follows that

$$E_n(0) = -\frac{1}{2}\sum_{k=0}^{n-1}\binom{n}{k}E_k(0),$$

but $E_n(0) = a_n$. Therefore, we have a recurrence relation for a_n:

$$a_{n,0} = a_n = -\frac{1}{2}\sum_{k=0}^{n-1}\binom{n}{k}a_k. \tag{8.100}$$

From (8.99), for $n > 0$ and $x = 0$, we have

$$\frac{1}{2} = \tilde{E}_n(0) = -\sum_{k=0}^{n-1}\binom{n}{k}a_{n-k}\tilde{E}_k(0) = -\frac{1}{2}\sum_{k=0}^{n-1}a_{n,k},$$

that is,

$$1 = -\sum_{k=0}^{n-1}a_{n,k}.$$

Therefore,

$$a_{n,0} = -\sum_{k=0}^{n}a_{n,k},$$

that is,

$$E_n(0) = -E_n(1). \tag{8.101}$$

From (8.96) and (8.97), it follows that

$$E_{2n}(1) = 0, \quad E_{2n-1}(1) = \frac{2^{2n}-1}{n}B_{2n}. \tag{8.102}$$

For $x = \frac{1}{2}$ in the generating function, we have

$$\frac{2\exp(\frac{t}{2})}{\exp(t)+1} = \frac{1}{\cosh\frac{1}{2}t} = \operatorname{sech}\frac{1}{2}t = \sum_{m=0}^{\infty}E_m\left(\frac{1}{2}\right)\frac{t^m}{m!}. \tag{8.103}$$

The numbers $2^n E_n(\frac{1}{2})$ are known as Euler numbers and denoted by E_n, that is,

$$E_n := 2^n E_n\left(\frac{1}{2}\right). \tag{8.104}$$

Moreover, from (8.103), we get

$$E_{2n+1}\left(\frac{1}{2}\right) = 0, \quad n \geq 0. \tag{8.105}$$

From (8.98), we have the first determinantal form for Euler polynomials

$$E_0(x) = 1,$$

$$E_n(x) = (-1)^n \begin{vmatrix} 1 & x & x^2 & \cdots & x^{n-1} & x^n \\ 1 & \frac{1}{2} & \frac{1}{2} & \cdots & \frac{1}{2} & \frac{1}{2} \\ 0 & 1 & \binom{2}{1}\frac{1}{2} & \cdots & \binom{n-1}{1}\frac{1}{2} & \binom{n}{1}\frac{1}{2} \\ \vdots & & \ddots & \ddots & & \\ 0 & & & \cdots & 1 & \binom{n}{n-1}\frac{1}{2} \end{vmatrix}. \tag{8.106}$$

For $x = 0$, the above determinant gives

$$a_n = E_n(0) = (-1)^n \begin{vmatrix} \frac{1}{2} & \frac{1}{2} & \cdots & \frac{1}{2} & \frac{1}{2} \\ 1 & \binom{2}{1}\frac{1}{2} & \cdots & \binom{n-1}{1}\frac{1}{2} & \binom{n}{1}\frac{1}{2} \\ \vdots & & \ddots & \ddots & \\ & & \cdots & 1 & \binom{n}{n-1}\frac{1}{2} \end{vmatrix} \tag{8.107}$$

which, combined with (8.100), gives a recurrent formula for the calculation of this determinant.

For the conjugate Euler polynomials, from (8.99), we have

$$\tilde{E}_0(x) = 1,$$

$$\tilde{E}_n(x) = (-1)^n \begin{vmatrix} 1 & x & x^2 & \cdots & x^{n-1} & x^n \\ 1 & E_1(0) & E_2(0) & \cdots & E_{n-1}(0) & E_n(0) \\ 0 & 1 & \binom{2}{1}E_1(0) & \cdots & \binom{n-1}{1}E_{n-2}(0) & \binom{n}{1}E_{n-1}(0) \\ \vdots & & \ddots & \ddots & & \\ 0 & & & \cdots & 1 & \binom{n}{n-1}E_1(0) \end{vmatrix}. \tag{8.108}$$

From Theorem (6.8), Euler polynomial $E_n(x)$ satisfies the differential equation

$$\frac{1}{2}\frac{1}{n!}y^{(n)}(x) + \frac{1}{2}\frac{1}{(n-1)!}y^{(n-1)}(x) + \cdots \frac{1}{2}y'(x) + y(x) = x^n. \tag{8.109}$$

For Euler polynomials, (6.36) becomes

$$b_i = \sum_{k=0}^{i} \binom{i}{k} \frac{E_{k+1}(0)}{2}. \tag{8.110}$$

Then we get the second recurrence relation (6.34) for Euler polynomials

$$E_{n+1}(x) = (x + b_0)E_n(x) + \sum_{k=0}^{n-1} \binom{n}{k} b_{n-k}E_k(x). \tag{8.111}$$

So for the Euler polynomials, we obtain the second determinantal form

$$E_{n+1}(x) = \begin{vmatrix} x+b_0 & -1 & & \cdots & 0 & 0 \\ b_1 & x+b_0 & -1 & \cdots & 0 & 0 \\ \vdots & & & \ddots & & \vdots \\ b_{n-1} & \binom{n-1}{1}b_{n-2} & \binom{n-1}{2}b_{n-3} & \cdots & x+b_0 & -1 \\ b_n & \binom{n}{1}b_{n-1} & \binom{n}{2}b_{n-2} & \cdots & \binom{n}{n-1}b_1 & x+b_0 \end{vmatrix}. \tag{8.112}$$

From (8.112), by applying Theorem 6.21, we have that the zeros of $E_n(x)$ lie inside the circle of center $b_0 = \frac{1}{2}$ and radius r, with

$$r = \max_{0\le i\le n} \sum_{\substack{j=0\\ i\ne j}}^{i+1} \binom{i}{k} \left| \sum_{k=0}^{i-j} \binom{i-j}{k} \frac{E_{k+1}(0)}{2} \right|. \tag{8.113}$$

In analogy, for the conjugate Euler polynomials, it results that

$$\tilde{b}_i = \sum_{k=0}^{i} \binom{i}{k} \frac{E_{i-k}(0)}{2}, \tag{8.114}$$

and hence we get the second recurrence relation

$$\hat{E}_{n+1}(x) = (x+\tilde{b}_0)\hat{E}_n(x) + \sum_{k=0}^{n-1} \binom{n}{k} \tilde{b}_{n-k}\hat{E}_k(x) \tag{8.115}$$

and the related determinantal form

$$\tilde{E}_{n+1}(x) = \begin{vmatrix} x+\tilde{b}_0 & -1 & & \cdots & 0 & 0 \\ \tilde{b}_1 & x+\tilde{b}_0 & -1 & \cdots & 0 & 0 \\ \vdots & & & \ddots & & \vdots \\ \tilde{b}_{n-1} & \binom{n-1}{1}\tilde{b}_{n-2} & \binom{n-1}{2}\tilde{b}_{n-3} & \cdots & x+\tilde{b}_0 & -1 \\ \tilde{b}_n & \binom{n}{1}\tilde{b}_{n-1} & \binom{n}{2}\tilde{b}_{n-2} & \cdots & \binom{n}{n-1}\tilde{b}_1 & x+\tilde{b}_0 \end{vmatrix}. \tag{8.116}$$

From (8.116), by applying Theorem 6.21, we have that the zeros of $\tilde{E}_n(x)$ lie inside the circle of center $\tilde{b}_0 = \frac{1}{2}$ and radius $\tilde{r}$, with

$$\tilde{r} = \max_{0\le i\le n} \sum_{\substack{j=0\\ i\ne j}}^{i+1} \binom{i}{k} \left| \sum_{k=0}^{i-j} \binom{i-j}{k} \frac{E_{i-j-k}(0)}{2} \right|. \tag{8.117}$$

Let $p_n(x)$ be the polynomial

$$p_n(x) = \sum_{k=0}^{n} c_{n,k}x^k.$$

From (8.98) and (8.99), we get expressions of $p_n(x)$ in the bases $\{E_0(x), E_1(x), \dots, E_n(x)\}$ and $\{\tilde{E}_0(x), \tilde{E}_1(x), \dots, \tilde{E}_n(x)\}$. We have, respectively,

$$p_n(x) = \sum_{k=0}^{n} c_{n,k}\left(E_k(x) + \frac{1}{2}\sum_{i=0}^{k-1}\binom{k}{i}E_i(x)\right) = \sum_{k=0}^{n}\left(c_{n,k} + \frac{1}{2}\sum_{i=k+1}^{n}\binom{i}{k}c_{n,i}\right)E_k(x),$$

$$p_n(x) = \sum_{k=0}^{n} c_{n,k}\left(\tilde{E}_k(x) + \sum_{i=0}^{k-1} a_{k,i}\tilde{E}_i(x)\right) = \sum_{k=0}^{n}\left(c_{n,k} + \sum_{i=k+1}^{n} a_{i,k}c_{n,i}\right)\tilde{E}_k(x).$$

Likewise, for a basis $\{q_k\}_k$, if

$$x^n = \sum_{k=0}^{n} q_{n,k}q_k(x),$$

we have

$$E_n(x) = \sum_{k=0}^{n}\sum_{i=k}^{n} a_{n,i}q_{i,k}q_k(x),$$

$$\tilde{E}_n(x) = 1 + \sum_{k=1}^{n}\sum_{i=k}^{n}\frac{1}{2}\binom{n}{i}q_{i,k}q_k(x).$$

Binomial identities for Euler and Euler conjugate polynomials are

$$E_n(x+y) = \sum_{k=0}^{n}\binom{n}{k}E_k(x)y^{n-k}, \quad \tilde{E}_n(x+y) = \sum_{k=0}^{n}\binom{n}{k}\tilde{E}_k(x)y^{n-k},$$

and for $y = 1$, we have

$$E_n(x+1) = \sum_{k=0}^{n}\binom{n}{k}E_k(x), \quad \tilde{E}_n(x+1) = \sum_{k=0}^{n}\binom{n}{k}\tilde{E}_k(x).$$

Now directly from (8.98), we have

$$x^n = E_n(x) + \frac{1}{2}\sum_{k=0}^{n-1}\binom{n}{k}E_k(x) = \frac{1}{2}\left(E_n(x) + \sum_{k=0}^{n}\binom{n}{k}E_k(x)\right) = \frac{1}{2}(E_n(x) + E_n(x+1)),$$

which can be rewritten as

$$\mathcal{M}E_n(x) = x^n, \tag{8.118}$$

where $\mathcal{M}$ is the operator of mean [113, p. 6].

We consider now $\mathcal{M}E_{2n+1}(x) = x^{2n+1}$, and by setting $x = -z$, we get

$$\mathcal{M}E_{2n+1}(-z) = -z^{2n+1} = -\mathcal{M}E_{2n+1}(z). \tag{8.119}$$

On the other hand, we have

$$\mathcal{M}E_{2n+1}(x) = \frac{1}{2}[E_{2n+1}(x+1) + E_{2n+1}(x)].$$

If $x = -z$, we obtain

$$\mathcal{M}E_{2n+1}(-z) = \mathcal{M}E_{2n+1}(1-z), \tag{8.120}$$

or, in general,

$$\mathcal{M}E_{2n+1}(x) = -\mathcal{M}E_{2n+1}(1-x). \tag{8.121}$$

Then by applying the operator $\mathcal{M}^{-1}$ [113, p. 297], we have

$$E_{2n+1}(x) = -E_{2n+1}(1-x)$$

or

$$E_{2n+1}\left(\frac{1}{2}+x\right) = -E_{2n+1}\left(\frac{1}{2}-x\right). \tag{8.122}$$

From (8.122), for $x = 0$, according to (8.105), it follows that

$$E_{2n+1}\left(\frac{1}{2}\right) = 0. \tag{8.123}$$

By the differentiation of (8.122), we have

$$\begin{aligned} E_{2n}(x) &= E_{2n}(1-x), \\ E_{2n}\left(\frac{1}{2}+x\right) &= E_{2n}\left(\frac{1}{2}-x\right), \end{aligned} \tag{8.124}$$

For $x = \frac{1}{2}$ in (8.124), we get

$$E_{2n}(1) = E_{2n}(0),$$

but we have already seen that

$$E_{2n}(0) = -E_{2n}(1)$$

for $n > 1$. Hence

$$\forall n > 1, \quad E_{2n}(0) = E_{2n}(1) = 0.$$

Consequently, we say that $E_{2n}(x)$ is symmetric with respect to $x = \frac{1}{2}$, and it is divisible by $x(x-1)$. Moreover, for $x = \frac{1}{2}$, it has one extreme point.

For the roots of Euler polynomials, there is a wide literature; for example, see [106, 107] and references therein.

For the conjugate Euler polynomial, we have

$$\tilde{E}_{2n}(x) > 0 \quad \forall x \in \mathbb{R}, \quad \tilde{E}_{2n+1}(x) > 0 \quad \forall x \geq 0.$$

Moreover, $\tilde{E}_{2n}(x)$ has an extreme point for $x = -\frac{1}{2}$. In fact, $\tilde{E}_{2n+1}(x)$ has only the real root $x = -\frac{1}{2}$.

Let X be a linear space for real functions, and let the linear functionals on X be defined by

$$L(f) = (\mathcal{M}f)_{x=0}, \quad L_1(f) = (\mathcal{M}^{-1}f)_{x=0}. \tag{8.125}$$

We consider the polynomials

$$P_n[f](x) = \sum_{i=0}^{n} L(f^{(i)}) \frac{E_i(x)}{i!}, \tag{8.126}$$

$$Q_n[f](x) \sum_{i=0}^{n} L_1(f^{(i)}) \frac{\tilde{E}_i(x)}{i!}. \tag{8.127}$$

From Theorem 7.1, they satisfy the interpolant conditions

$$L(f^{(i)}) = L(P_n^{(i)}[f]), \quad i = 0, \dots, n, \tag{8.128}$$

$$L_1(f^{(i)}) = L(Q_n^{(i)}[f]), \quad i = 0, \dots, n. \tag{8.129}$$

The remainder term can be calculated by Theorem 7.3.

The polynomial (8.126) is very interesting for interpolation. Therefore, we rewrite it as

$$P_n[f](x) = \sum_{i=0}^{n} \frac{f^{(i)}(1) + f^{(i)}(0)}{2} \frac{E_i(x)}{i!}. \tag{8.130}$$

Conjugate Euler polynomials, such as conjugate Bernoulli polynomials, are more important for approximation of real function on a semiinfinite interval.

In fact, denoting by E the class of functions of exponential type, that is, functions with the property that $|f(t)| \leq e^{At}$ for each $t \geq 0$ and some finite number A, we have the following:

Theorem 8.4. *If $f \in C[0, \infty[\cap E$ and*

$$P_n(f)(x) = \frac{2\exp(-nx)}{\exp(1) + 1} \sum_{k=0}^{\infty} \tilde{E}_k(nx) f\left(\frac{k}{n}\right), \tag{8.131}$$

we have

$$\lim_{n \to \infty} P_n(f)(x) = f(x) \quad \forall x \in [0, +\infty[, \tag{8.132}$$

and the convergence is uniform in each compact $[0, a]$, $a > 0$.

For conjugate Euler polynomials we can also apply the extrapolation procedure of Theorem 7.8, which provides a faster convergence [55].

8.3 Hermite polynomials

Hermite polynomials were defined by Laplace [115, 122] and studied in detail by Chebyshev [35]. Chebyshev's work was overlooked, and Hermite polynomials were considered in 1864 by Charles Hermite, who described them as new. They were consequently not new, although in later 1865 papers, Hermite was the first to define the multidimensional polynomials.

The Hermite polynomials have many applications, for example, probability (probabilistic Hermite polynomials denoted by $\mathrm{He}_n(x)$), combinatorics, numerical analysis (finite element methods), physics (physicist Hermite polynomials denoted by $H_n(x)$). As in the cases of Bernoulli and Euler polynomials, we now give a brief profile of the physicists' Hermite polynomials.

We consider the numerical sequence $(\widehat{a}_i)_i$, defined as

$$\begin{cases} \widehat{a}_0 = 1, & \\ \widehat{a}_{2i+1} = 0, & i = 0, 1, \dots, \\ \widehat{a}_{2i} = \frac{(2i-1)(2i-3)\cdots 3\cdot 1}{2^i} = \frac{(2i-1)!!}{2^i}, & i = 1, 2, 3 \dots, \end{cases} \tag{8.133}$$

and the matrix $\widehat{A} := (\widehat{a}_{i,j})$, $i = 0, 1, \dots, j = 0, \dots, i$, with

$$\widehat{a}_{i,j} = \begin{cases} \binom{i}{j}\widehat{a}_{i-j}, & j \le i, \\ 0, & j > i. \end{cases} \tag{8.134}$$

Given that $\widehat{a}_{i,i} = 1$, we can consider the matrix $A = \widehat{A}^{-1}$, and if $A := (a_{i,j})$, we have

$$a_{i,j} = \begin{cases} \binom{i}{j}a_{i-j}, & j \le i, \\ 0, & j > i, \end{cases} \tag{8.135}$$

where the sequence $(a_i)_i$ satisfies

$$\forall i \ge 0, \quad \sum_{k=0}^{i} \binom{i}{k} a_k \widehat{a}_{i-k} = \delta_{i,0}. \tag{8.136}$$

Remark 8.1. From (8.133) and (8.136), we have

$$\begin{cases} a_{2i+1} = 0, & i = 0, 1, \dots, \\ a_{2i} = (-1)^i \frac{(2i-1)!!}{2^i} = (-1)^i \widehat{a}_{2i} & i = 1, 2, \dots. \end{cases} \tag{8.137}$$

Then we get the conjugate A.p.s:

$$
\begin{aligned}
&a_0(x) = a_{0,0}, && \widehat{a}_0(x) = \widehat{a}_{0,0},\\
&a_1(x) = a_{1,0} + a_{1,1}x, && \widehat{a}_1(x) = \widehat{a}_{1,0} + \widehat{a}_{1,1}x,\\
&\cdots\cdots\cdots && \cdots\cdots\cdots \\
&a_n(x) = a_{n,0} + a_{n,1}x + \cdots + a_{n,n}x^n, && \widehat{a}_n(x) = \widehat{a}_{n,0} + \widehat{a}_{n,1}x + \cdots + \widehat{a}_{n,n}x^n,\\
&\cdots\cdots\cdots && \cdots\cdots\cdots
\end{aligned}
\tag{8.138}
$$

The first sequence $\{a_n\}_n$ is known as physicists', normalized Hermite polynomials and denoted by $\{H_n\}_n$, whereas we call the second sequence $\{\widehat{a}_n\}_n$ conjugate physicists' Hermite polynomials and denote it by $\{\tilde{H}_n\}_n$.

In Table 8.4, the first six polynomials of these sequences are listed.

Then in Figures 8.5 and 8.6, we compare the graphics of the first six Hermite and conjugate Hermite polynomials.

Directly from (8.133) and (8.138), we have

$$
H_n(x) = \sum_{k=0}^{n} \binom{n}{k} H_{n-k}(0)x^n = \sum_{k=0}^{[\frac{n}{2}]} \binom{n}{2k} (-1)^k \frac{(2k-1)!!}{2^k} x^{n-2k}, \tag{8.139}
$$

Table 8.4: Hermite and Hermite conjugate polynomials.

Hermite polynomials	**Hermite conjugate polynomials**
$H_0(x) = 1$	$\tilde{H}_0(x) = 1$
$H_1(x) = x$	$\tilde{H}_1(x) = x$
$H_2(x) = x^2 - \frac{1}{2}$	$\tilde{H}_2(x) = x^2 + \frac{1}{2}$
$H_3(x) = x^3 - \frac{3}{2}x$	$\tilde{H}_3(x) = x^3 + \frac{3}{2}x$
$H_4(x) = x^4 - 3x^2 + \frac{3}{4}$	$\tilde{H}_4(x) = x^4 + 3x^2 + \frac{3}{4}$
$H_5(x) = x^5 - 5x^3 + \frac{15}{4}x$	$\tilde{H}_5 = x^5 + 5x^3 + \frac{15}{4}x$

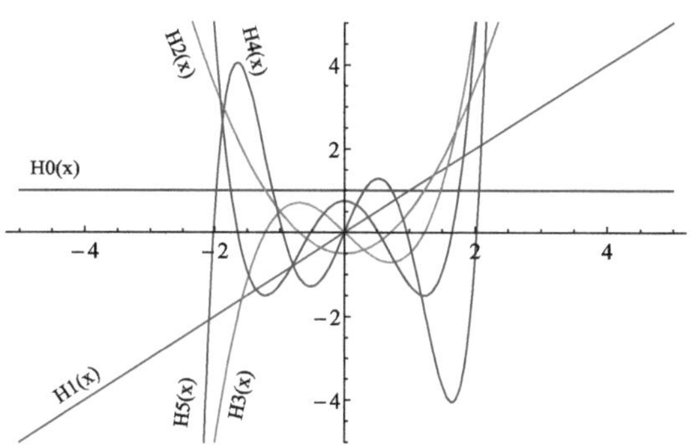

Figure 8.5: Hermite polynomials.

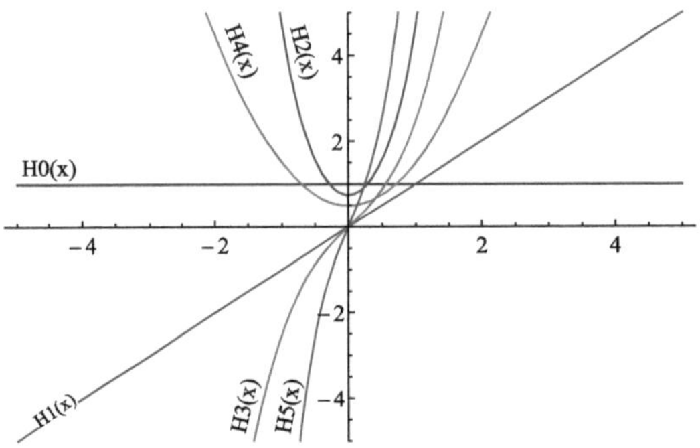

Figure 8.6: Hermite conjugate polynomials.

where $(2k-1)!! = 1 \cdot 3 \cdot 5 \cdots (2k-1)$, and

$$\tilde{H}_n(x) = \sum_{k=0}^{n} \binom{n}{k} \tilde{H}_{n-k}(0) x^n = \sum_{k=0}^{[\frac{n}{2}]} \binom{n}{2k} \frac{(2k-1)!!}{2^k} x^{n-2k}. \tag{8.140}$$

So we have

$$H_n(\tilde{H}_n(x)) = \tilde{H}_n(H_n(x)) = x^n. \tag{8.141}$$

Theorem 6.1 results in

$$H_n'(x) = nH_{n-1}(x), \quad \tilde{H}_n'(x) = n\tilde{H}_{n-1}(x), \tag{8.142}$$

and for integration, we get

$$\int_0^x H_n(t)dt = \frac{1}{n+1}(H_{n+1}(x) - H_{n+1}(0)), \quad \int_0^x \tilde{H}_n(t)dt = \frac{1}{n+1}(\tilde{H}_{n+1}(x) - \tilde{H}_{n+1}(0)). \tag{8.143}$$

So for the definite integral, we have

$$\int_0^1 H_n(x)dx = \begin{cases} \frac{1}{n+1} \sum_{k=0}^{\frac{n-1}{2}} \binom{n+1}{2k} (-1)^k \frac{(2k-1)!!}{2^k}, & n \text{ odd}, \\ \frac{1}{n+1} \sum_{k=0}^{\frac{n-2}{2}} \binom{n+1}{2k} (-1)^k \frac{(2k-1)!!}{2^k}, & n \text{ even}, \end{cases} \tag{8.144}$$

$$\int_0^1 \tilde{H}_n(x)dx = \begin{cases} \frac{1}{n+1} \sum_{k=0}^{\frac{n-1}{2}} \binom{n+1}{2k} \frac{(2k-1)!!}{2^k}, & n \text{ odd}, \\ \frac{1}{n+1} \sum_{k=0}^{\frac{n-2}{2}} \binom{n+1}{2k} \frac{(2k-1)!!}{2^k}, & n \text{ even}. \end{cases} \tag{8.145}$$

Consequently, the operational matrix of integration is

$$\mathcal{I} = \overline{U}\hat{A}_n, \tag{8.146}$$

where

$$\overline{U} = [U_1, \ldots, U_n, W^T A_n]^T,$$

given that

$$U_i = \left[0, \frac{1}{i}\binom{i}{i-1}H_{i-1}(0), \dots, \frac{1}{i}\binom{i}{0}H_0(0), \underbrace{0, \dots, 0}_{n-i}\right],$$

and $W = [\mathcal{X}_0, \dots, \mathcal{X}_n]$ in

$$\frac{H_{n+1}(x) - H_{n+1}(0)}{n+1} \simeq \sum_{i=0}^{n} \mathcal{X}_i H_i(x).$$

For the conjugate Hermite polynomials, we have that

$$\mathcal{I} = \overline{U} A_n,$$

where

$$\overline{U} = [U_1, \dots, U_n, W^T \widehat{A}_n]^T,$$

given that

$$U_i = \left[0, \frac{1}{i}\binom{i}{i-1}\tilde{H}_{i-1}(0), \dots, \frac{1}{i}\binom{i}{0}\tilde{H}_0(0), \underbrace{0, \dots, 0}_{n-i}\right],$$

and $W = [\mathcal{X}_0, \dots, \mathcal{X}_n]$ in

$$\frac{\tilde{H}_{n+1}(x) - \tilde{H}_{n+1}(0)}{n+1} \simeq \sum_{i=0}^{n} \mathcal{X}_i \tilde{H}_i(x).$$

For

$$\widehat{a}(t) = \sum_{n=0}^{\infty} \widehat{a}_n \frac{t^n}{n!}$$

with $(\widehat{a}_i)$ as in (8.133), after calculation we have

$$\widehat{a}(t) = \exp\left(\frac{t^2}{4}\right). \tag{8.147}$$

Given that $a(t)\widehat{a}(t) = 1$, we have

$$a(t) = \exp\left(-\frac{t^2}{4}\right). \tag{8.148}$$

Consequently, the generating functions of Hermite and Hermite conjugate polynomials are

$$G(x,t) = \exp\left(xt - \frac{t^2}{4}\right), \quad \tilde{G}(x,t) = \exp\left(xt + \frac{t^2}{4}\right), \tag{8.149}$$

that is,

$$\exp\left(xt - \frac{t^2}{4}\right) = \sum_{n=0}^{\infty} H_n(x)\frac{t^n}{n!}, \quad \exp\left(xt + \frac{t^2}{4}\right) = \sum_{n=0}^{\infty} \tilde{H}_n(x)\frac{t^n}{n!}.$$

Directly from (6.26), we have the recurrence relations

$$H_n(x) = x^n - \sum_{k=0}^{[\frac{n-1}{2}]} \binom{n}{2k} \frac{(2(n-k))!!}{2^{n-k}} H_{2k}(x), \tag{8.150}$$

$$\tilde{H}_n(x) = x^n - \sum_{k=0}^{[\frac{n-1}{2}]} \binom{n}{2k} (-1)^{n-k} \frac{(2(n-k))!!}{2^{n-k}} \tilde{H}_{2k}(x). \tag{8.151}$$

The recurrence formula (8.150) is equivalent to the following determinantal form :

$$H_0(x) = 1,$$

$$H_n(x) = (-1)^n \begin{vmatrix} 1 & x & x^2 & \cdots & \cdots & x^{n-1} & x^n \\ 1 & 0 & \hat{a}_2 & \cdots & \cdots & \hat{a}_{n-1} & \hat{a}_n \\ 0 & 1 & 0 & \cdots & \cdots & \binom{n}{n-2}\hat{a}_{n-2} & \binom{n}{n-1}\hat{a}_{n-1} \\ \vdots & & \ddots & & & & \vdots \\ \vdots & & & & \ddots & & \vdots \\ 0 & & & & & 1 & 0 \end{vmatrix}, \tag{8.152}$$

where $\hat{a}_i$ are defined in (8.133).

For the conjugate Hermite polynomials, we have

$$\tilde{H}_0(x) = 1,$$

$$\tilde{H}_n(x) = (-1)^n \begin{vmatrix} 1 & x & x^2 & \cdots & \cdots & x^{n-1} & x^n \\ 1 & 0 & a_2 & \cdots & \cdots & a_{n-1} & a_n \\ 0 & 1 & 0 & \cdots & \cdots & \binom{n}{n-2}a_{n-1} & \binom{n}{n-1}a_{n-1} \\ \vdots & & \ddots & & & & \vdots \\ \vdots & & & & \ddots & & \vdots \\ 0 & & & & & 1 & 0 \end{vmatrix}, \tag{8.153}$$

where a_i are defined in (8.137).

From (6.36), we get the sequence $(b_i)_i$ such that

$$b_0 = 0, \quad b_1 = -\frac{1}{2}, \quad b_i = 0, \quad i = 2, \dots. \tag{8.154}$$

From (6.34), after easy calculations, we get the second recurrence relation for Hermite polynomials

$$H_{n+1}(x) = xH_n(x) - \frac{1}{2}nH_{n-1}(x). \tag{8.155}$$

So for Hermite polynomials, we get the second determinantal form

$$H_{n+1}(x) = \begin{vmatrix} x & -1 & 0 & & \cdots & 0 \\ -\frac{1}{2} & x & -1 & & \cdots & 0 \\ 0 & -\binom{2}{1}\frac{1}{2} & x & -1 & \cdots & 0 \\ \vdots & & & \ddots & & \vdots \\ 0 & 0 & 0 & \cdots & \cdots & -1 \\ 0 & 0 & 0 & \cdots & -\binom{n}{n-1}\frac{1}{2} & x \end{vmatrix}. \tag{8.156}$$

Analogously for the conjugate Hermite polynomials, defining $(\tilde{b}_i)_i$ as

$$\tilde{b}_0 = 0, \quad \tilde{b}_i = \frac{1}{2}, \quad \tilde{b}_i = 0, \quad i = 2, \dots, \tag{8.157}$$

we get the second recurrence relation

$$\tilde{H}_{n+1}(x) = x\tilde{H}_n(x) + \frac{1}{2}n\tilde{H}_{n-1}(x) \tag{8.158}$$

and the related determinantal form

$$\tilde{H}_{n+1}(x) = \begin{vmatrix} x & -1 & 0 & & \cdots & 0 \\ \frac{1}{2} & x & -1 & & \cdots & 0 \\ 0 & \binom{2}{1}\frac{1}{2} & x & -1 & \cdots & 0 \\ \vdots & & & \ddots & & \vdots \\ 0 & 0 & 0 & \cdots & \cdots & -1 \\ 0 & 0 & 0 & \cdots & \binom{n}{n-1}\frac{1}{2} & x \end{vmatrix}. \tag{8.159}$$

We note explicitly that the three-term recurrence relation (8.158) does not imply the orthogonality of conjugate Hermite sequence. In fact, it does not verify the hypothesis of Favard's theorem [90]. To the contrary, the relation (8.154) implies the orthogonality of the Hermite polynomial sequence $\{H_n\}_n$.

The first differential equation ([213, 215]) is not very important, whereas the second one gives the classical equation

$$H_n''(x) - 2xH_n'(x) + 2nH_n(x) = 0. \tag{8.160}$$

This differential equation allows us to prove the orthogonality property [163, p. 192–193]

$$\int_{-\infty}^{\infty} exp(-x^2)H_n(x)H_m(x)dx = 0, \quad n \neq m, \quad n, m \in \mathbb{N}. \tag{8.161}$$

Moreover, the Hermite polynomials are the only A. p. s. that are also classic orthogonal [185]. Consequently, the zeros of Hermite polynomials are all real and simple.

For linear interpolation, we observe that for the linear functional

$$L(f) = \frac{1}{\sqrt{\pi}} \int_{-\infty}^{+\infty} \exp(-x^2) f(x)dx, \tag{8.162}$$

we get

$$L(x^i) = \frac{1}{\sqrt{\pi}} \int_{-\infty}^{+\infty} \exp(-x^2)x^i dx = \hat{a}_i,$$

and polynomials $a_{L,n}(x)$ coincide with Hermite polynomials, that is,

$$a_{L,n}(x) = H_n(x). \tag{8.163}$$

The interpolant polynomial is

$$H_n[f](x) = \frac{1}{\sqrt{\pi}} \sum_{i=0}^{n} \frac{1}{i!} H_i(x) \int_{-\infty}^{+\infty} \exp(-x^2) f^{(i)}(x)dx. \tag{8.164}$$

From the Peano lemma, the remainder is

$$R_{L,n}[f](x) = \int_{0}^{1} K(x,t) f^{(n+1)}(t)dt, \tag{8.165}$$

where

$$K(x,t) = (x-t)_+^n - \frac{1}{\sqrt{\pi}} \sum_{i=0}^{n} \frac{1}{i!} H_i(x) \int_{-\infty}^{\infty} \exp(-x^2)[(x-t)_+^n]^{(i)} dx.$$

The conjugate Hermite polynomials are more important for approximation of real functions on a semiinfinite interval.

In fact, denoting by E the class of all functions of exponential type, that is, functions with the property $|f(t)| \le e^{At}$ for each $t \ge 0$ and some finite number A, we have the following:

Theorem 8.5. *If $f \in C[0,\infty[\cap E$, the linear functional sequence*

$$P_n(f)(x) = \frac{\exp(-nx)}{\exp(-\frac{1}{4})} \sum_{k=0}^{\infty} \tilde{H}_k(nx) f\left(\frac{k}{n}\right) \tag{8.166}$$

converges uniformly to f in each compact of type $[0,A]$; that is,

$$\lim_{n\to\infty} P_n(f)(x) = f(x) \quad \forall x \in [0,+\infty[, \tag{8.167}$$

and the convergence is uniform in each compact $[0,A]$, $A > 0$.

For conjugate Hermite polynomials, we can apply the extrapolation procedure of Theorem 7.8, which provides a faster convergence.

8.4 Appell–Laguerre polynomials

If we consider the sequence $(\widehat{a}_i)_i$:

$$\widehat{a}_i = i!, \quad i = 0, 1, \dots, \tag{8.168}$$

we get the Appell matrix $\widehat{A} := (\widehat{a}_{i,j})$ with

$$\widehat{a}_{i,j} = \binom{i}{j}(i-j)! = \frac{i!}{j!}. \tag{8.169}$$

Then we obtain the conjugate sequence $(a_i)_i$ with

$$a_0 = 1, \quad a_1 = -1, \quad a_i = 0, \quad i = 2, 3, \dots, \tag{8.170}$$

and consequently the matrix $A := (a_{i,j})$ with

$$\begin{cases} a_{0,0} = 1, \\ a_{i,i} = 1, \\ a_{i,i-1} = -i, \\ a_{i,j} = 0 \qquad \text{otherwise}. \end{cases} \tag{8.171}$$

Given A and $\widehat{A}$ Appell matrices, we can define the A. p. s. $\{a_n\}_n$ and $\{\widehat{a}_n\}_n$:

$$\begin{array}{ll} a_0(x) = 1, & \widehat{a}_0(x) = 1, \\ a_1(x) = a_{1,0} + a_{1,1}x, & \widehat{a}_1(x) = \widehat{a}_{1,0} + \widehat{a}_{1,1}x, \\ \cdots\cdots\cdots & \cdots\cdots\cdots \\ a_n(x) = a_{n,0} + a_{n,1}x + \cdots + a_{n,n}x^n, & \widehat{a}_n(x) = \widehat{a}_{n,0} + \widehat{a}_{n,1}x + \cdots + \widehat{a}_{n,n}x^n, \\ \cdots\cdots\cdots & \cdots\cdots\cdots \end{array}$$

Observing that

$$\widehat{a}_i = \int_0^\infty \exp(-x)x^i dx, \tag{8.172}$$

we can call the sequences $\{a_n\}_n$ and $\{\widehat{a}_i\}_i$ conjugate Appell–Laguerre polynomial sequences.

In Table 8.5, the first six conjugate Appell–Laguerre polynomials are listed.

In Figures 8.7 and 8.8 we display their graphics.

If we consider the power series

$$a(t) = \sum_{n=0}^\infty a_n \frac{t^n}{n!}, \quad \widehat{a}(t) = \sum_{n=0}^\infty \widehat{a}_n \frac{t^n}{n!},$$

Table 8.5: Appell–Laguerre and conjugate Appell–Laguerre polynomials.

Appell–Laguerre polynomials	Appell–Laguerre conjugate polynomials
$a_0(x) = 1$	$\hat{a}_0(x) = 1$
$a_1(x) = x - 1$	$\hat{a}_1(x) = x + 1$
$a_2(x) = x^2 - 2x$	$\hat{a}_2(x) = x^2 + 2x + 2$
$a_3(x) = x^3 - 3x^2$	$\hat{a}_3(x) = x^3 + 3x^2 + 6x + 6$
$a_4(x) = x^4 - 4x^3$	$\hat{a}_4(x) = x^4 + 4x^3 + 12x^2 + 24x + 24$
$a_5(x) = x^5 - 5x^4$	$\hat{a}_1(x) = x^5 + 5x^4 + 20x^3 + 60x^2 + 120x + 120$

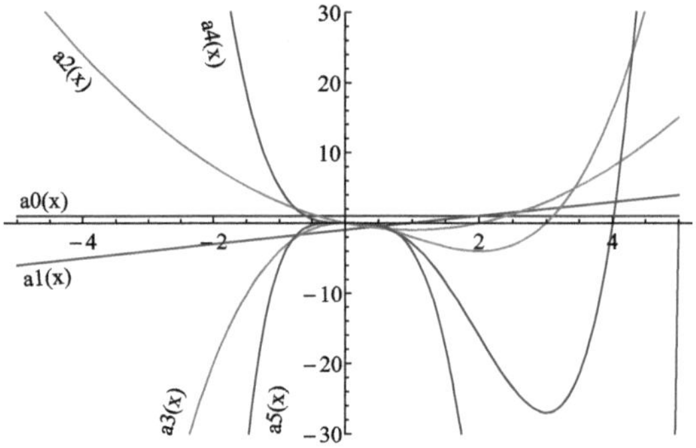

Figure 8.7: Appell–Laguerre polynomials.

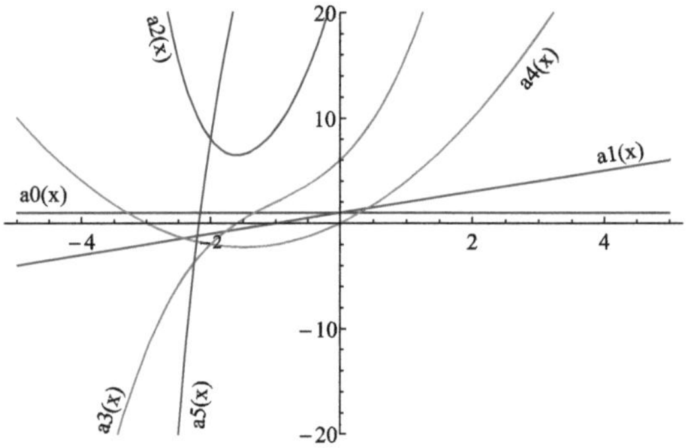

Figure 8.8: Appell–Laguerre conjugate polynomials.

we have

$$\hat{a}(t) = \sum_{n=1}^{\infty} t^n = \frac{1}{1-t}, \tag{8.173}$$

whereas for $a(t)$, given that $a(t)\hat{a}(t) = 1$, we have

$$a(t) = 1 - t. \tag{8.174}$$

Consequently,

$$G(x,t) = (1-t)\exp(xt), \quad \tilde{G}(x,t) = \frac{1}{1-t}\exp(xt) \tag{8.175}$$

are the generating functions of $\{a_n\}$ and $\{\widehat{a}_n\}$. In fact, we have

$$(1-t)\exp(xt) = \sum_{n=0}^{\infty} a_n(x)\frac{t^n}{n!}, \quad \frac{1}{1-t}\exp(xt) = \sum_{n=0}^{\infty} \widehat{a}_n(x)\frac{t^n}{n!}. \tag{8.176}$$

Directly from (6.26), we have the recurrence relations

$$a_n(x) = x^n - \sum_{k=0}^{n-1} \frac{n!}{k!} a_k(x), \tag{8.177}$$

$$\widehat{a}_n(x) = x^n + n\widehat{a}_{n-1}(x). \tag{8.178}$$

The recurrence formula (8.177) is equivalent to the following determinantal form ([60]):

$$a_0(x) = 1,$$

$$a_n(x) = (-1)^n \begin{vmatrix} 1 & x & x^2 & x^3 & \cdots & x^{n-1} & x^n \\ 1 & 1 & 2! & 3! & \cdots & (n-1)! & n! \\ 0 & 1 & \frac{2!}{1!} & \frac{3!}{1!} & \cdots & \frac{(n-2)!}{1!} & \frac{n!}{1!} \\ 0 & 0 & 1 & \frac{3!}{2!} & \cdots & \frac{(n-2)!}{2!} & \frac{n!}{2!} \\ \vdots & & & & \ddots & & \vdots \\ 0 & & & & & 1 & \frac{n!}{(n-1)!} \end{vmatrix} = x^{n-1}(x-n). \tag{8.179}$$

For the conjugate Appell–Laguerre polynomials, we have

$$\widehat{a}_0(x) = 1,$$

$$\widehat{a}_n(x) = (-1)^n \begin{vmatrix} 1 & x & x^2 & x^3 & \cdots & x^{n-1} & x^n \\ 1 & -1 & 0 & 0 & \cdots & 0 & 0 \\ 0 & 1 & -2 & 0 & \cdots & 0 & 0 \\ 0 & 0 & 1 & -3 & \cdots & 0 & 0 \\ \vdots & & & & \ddots & & \vdots \\ 0 & & & & & 1 & -n \end{vmatrix}. \tag{8.180}$$

Directly from (6.36), we get

$$b_i = -i!, \quad i = 0\,1, \ldots. \tag{8.181}$$

Therefore, the second recurrence relation (6.34) for Appell–Laguerre polynomials is

$$a_{n+1}(x) = (x-1)a_n(x) - \sum_{k=0}^{n-1} \frac{n!}{k!} a_k(x). \tag{8.182}$$

So for Appell–Laguerre polynomials, we get the second determinantal form

$$a_{n+1}(x) = \begin{vmatrix} x-1 & -1 & 0 & & \cdots & & 0 \\ -1 & x-1 & -1 & & \cdots & & 0 \\ -2! & -2! & x-1 & -1 & \cdots & & 0 \\ \vdots & & & \ddots & & & \vdots \\ -(n-1)! & -(n-1)! & -\frac{(n-1)!}{2!} & & \cdots & x-1 & -1 \\ -n! & -n! & -\frac{n!}{2!} & & \cdots & & x-1 \end{vmatrix}. \tag{8.183}$$

Likewise, for the conjugate Appell–Laguerre polynomials, we get

$$\tilde{b}_0 = 1, \quad \tilde{b}_i = i!i, \quad i = 1, \dots. \tag{8.184}$$

Then the second recurrence relation is

$$\widehat{a}_{n+1}(x) = (x+1)\widehat{a}_n(x) + \sum_{k=0}^{n-1} \frac{n!}{k!}(n-k)\widehat{a}_k(x), \tag{8.185}$$

and the related determinantal form is

$$\widehat{a}_{n+1}(x) = \begin{vmatrix} x+\tilde{b}_0 & -1 & 0 & & \cdots & & 0 \\ \tilde{b}_1 & x+\tilde{b}_0 & -1 & & \cdots & & 0 \\ \tilde{b}_2 & \binom{2}{1}\tilde{b}_1 & x+\tilde{b}_0 & -1 & \cdots & & 0 \\ \vdots & & & \ddots & & & \vdots \\ \tilde{b}_{n-1} & \binom{n-1}{1}\tilde{b}_{n-2} & \binom{n-1}{2}\tilde{b}_{n-3} & & \cdots & x+\tilde{b}_0 & -1 \\ \tilde{b}_n & \binom{n}{1}\tilde{b}_{n-1} & \binom{n}{2}\tilde{b}_{n-2} & & \cdots & & x+\tilde{b}_0 \end{vmatrix}. \tag{8.186}$$

From (8.186), by applying Theorem 6.21, we have that the zeros of $\widehat{a}_n(x)$ lie inside the circle of center $\tilde{b}_0 = 1$ and radius $\tilde{r}$, where

$$\tilde{r} = \max_{0 \le i \le n} \sum_{\substack{j=0 \\ i \ne j}}^{i+1} \frac{i!}{j!}(i-j). \tag{8.187}$$

For linear interpolation, we observe that for the linear functional

$$L(f) = \int_0^{+\infty} \exp(-x)f(x)dx, \tag{8.188}$$

according to (8.172), we get, for any $f \in C^n[0, +\infty[$,

$$f(x) = \int_0^{+\infty} \exp(-x)f(x)dx + \sum_{i=1}^{n} a_i(x) \int_0^{+\infty} \exp(-x)f^{(i)}(x)dx + R_n[f](x). \tag{8.189}$$

The remainder is

$$R_{L,n}[f](x) = \int_0^1 K(x,t) f^{(n+1)}(t)dt, \tag{8.190}$$

where

$$K(x,t) = (x-t)_+^n - \sum_{i=0}^{n} a_i(x) \int_0^{\infty} \exp(-x)\left[(x-t)_+^n\right]^{(i)} dx.$$

The conjugate Appell–Laguerre polynomials are not important for approximation of real functions on a semiinfinite interval.

In fact, the sequence $\{\widehat{a}_n\}_n$ is positive for $x \geq 0$, but $\widehat{a}(1)$ is not defined.

Part IV: **Sheffer polynomial sequences**

... And more generally, we can use other operators than $\frac{d}{dx}$ and Δ, to define futher sets. We thus obtain all polynomial sets of type zero (as we denote them)...
I. M. Sheffer (1939)

9 Sheffer polynomial sequence

Abstract: Sheffer polynomials are introduced by matrix calculus, following the method of previous chapters. Recurrence relations, determinantal forms, and differential equations are given. Finally, links to δ-functionals and operators are sketched.

9.1 Introduction

Sheffer [179] defines an A-type classification for a set of polynomials. In particular, the so-called set of type zero contains relevant sequences, such as the Laguerre, Hermite, Bernoulli, and other polynomials, which arise in numerous problems in applied mathematics, such as approximation theory, in several other branches of mathematics, as well as in physics, chemistry, and engineering. For a relevant bibliography until 2000, we refer to [79] and references therein. In the past, the Sheffer polynomials have been studied with different approaches; for example, in [177], the connection between Sheffer A-zero-type polynomials and Riordan arrays is sketched; in [103], the isomorphism between the Sheffer group and Riordan group is proved.

Sheffer theory [179] is based on formal power series and on particular operators; Steffensen [185] used the same approach. Afterwards, Mullin and Rota [142], Roman and Rota [167] and Roman [166], using the operator's method, gave a beautiful, but not so simple, theory of umbral calculus, including Sheffer polynomials. Recently, in [43, 44, 60, 64, 65, 67, 67, 215], the authors sketched a new theory of Sheffer A-type 0, based on elementary tools of linear algebra. A generalization of this idea has been applied to special polynomials of two variables related to Gould–Hopper polynomials in [117, 118], where properties and relations between Sheffer and so-called Gould–Sheffer polynomials are derived.

An alternative approach to Sheffer A-type 0, based on the topic of linear algebra, is in [3, 120, 121, 168, 170, 213, 215].

In the following, we extend to Sheffer polynomials the algebraic method of the previous chapters, based on elementary matrix calculus.

9.2 Definition and first characterizations

Let $a := (a_i)_i$ and $b := (b_i)_i$ be two sequences of elements of K with

$$a_0 \neq 0, \quad b_0 = 0, \quad b_1 \neq 0. \tag{9.1}$$

Then we can construct the following:
1. the conjugate sequences $\widehat{a} := (\widehat{a}_i)$ and $\widehat{b} := (\widehat{b}_i)$ as in (2.10) and (2.5);

https://doi.org/10.1515/9783110652925-009

2. the matrices $A := (a_{i,j})$ and $\widehat{A} := (\widehat{a}_{i,j})$ defined in Algorithm 2.2.1 and Algorithm 2.2.2; consequently, we get the conjugate A. p. s. $\{a_n\}_n$, $\{\widehat{a}_n\}_n$;
3. the matrices $P := (p_{i,j})$ and $\widehat{P} := (\widehat{p}_{i,j})$ as in Algorithm 2.1.1 and Algorithm 2.1.2; therefore, we have the conjugate b. p. s. $\{p_n\}_n$ and $\{\widehat{p}_n\}_n$;
4. the matrices $S := (s_{i,j})$ and $\widehat{S} := (\widehat{s}_{i,j})$ as in Algorithms 2.3.1 and 2.3.2 or equivalently by Proposition 2.4;
5. the δ-operator on $\mathcal{P}$:

$$Qy = \sum_{i=1}^{\infty} \widehat{b}_i \frac{y^{(i)}}{i!}, \quad \widehat{Q}y = \sum_{i=1}^{\infty} b_i \frac{y^{(i)}}{i!}, \tag{9.2}$$

with the properties

$$Qp_n(x) = np_{n-1}(x), \quad \widehat{Q}\widehat{p}_n(x) = n\widehat{p}_{n-1}(x), \tag{9.3}$$

where $\{p_n\}$ and $\{\widehat{p}_n\}$ are the b. p. s. related to the matrices P and $\widehat{P}$;
6. the formal power series

$$g(t) = \sum_{i=0}^{\infty} a_i \frac{t^i}{i!}, \quad \frac{1}{g(t)} = \sum_{i=0}^{\infty} \widehat{a}_i \frac{t^i}{i!}, \tag{9.4}$$

$$f(t) = \sum_{i=0}^{\infty} b_i \frac{t^i}{i!}, \quad \overline{f}(t) = \sum_{i=0}^{\infty} \widehat{b}_i \frac{t^i}{i!}, \tag{9.5}$$

with

$$f(\overline{f}(t)) = t$$

and

$$g(t) \cdot \frac{1}{g(t)} = 1.$$

Now we can define:

Definition 9.1. The polynomials sequences

$$\begin{aligned}
&s_0(x) = s_{0,0}, && \widehat{s}_0(x) = \widehat{s}_{0,0},\\
&s_1(x) = s_{1,0} + s_{1,1}x, && \widehat{s}_1(x) = \widehat{s}_{1,0} + \widehat{s}_{1,1}x,\\
&\cdots\cdots\cdots && \cdots\cdots\cdots\\
&s_n(x) = s_{n,0} + s_{n,1}x + \cdots + s_{n,n}x^n, && \widehat{s}_n(x) = \widehat{s}_{n,0} + \widehat{s}_{n,1}x + \cdots + \widehat{s}_{n,n}x^n,\\
&\cdots\cdots\cdots && \cdots\cdots\cdots
\end{aligned} \tag{9.6}$$

are named the *conjugate Sheffer sequences* for $((a_i), (b_i))_{i\in\mathbb{N}}$ or, for simplicity, the *conjugate Sheffer polynomial sequences* (in the following, C. S. p. s.) .

Proposition 9.1. *For any $n \in \mathbb{N}$ with umbral composition [167, p. 120], we have*

$$s_n(\widehat{s}_n(x)) = \widehat{s}_n(s_n(x)) = x^n, \tag{9.7}$$

that is,

$$\sum_{k=0}^{n} s_{n,k}\widehat{s}_k(x) = \sum_{k=0}^{n} \widehat{s}_{n,k}s_k(x) = x^n. \tag{9.8}$$

Proof. Relation (9.7) follows from Proposition 2.4, given that S and $\widehat{S}$ are the inverse matrices of each other. □

Proposition 9.2. *For any $n \in \mathbb{N}$, the polynomial sets $\{s_0(x), \dots, s_n(x)\}$ and $\{\widehat{s}_0(x), \dots, \widehat{s}_n(x)\}$ comprise a basis for $\mathcal{P}_n$.*

Proof. The result follows from Proposition 9.1. □

Proposition 9.2 allows us to change the basis from the canonical to a Sheffer one, as follows:

Theorem 9.1 (Representation theorem). *If $\{s_n\}_n$ is the S. p. s. for $((a_i), (b_i))$ and $q_n(x)$ is a polynomial of degree $\leq n$ such that $q_n(x) = \sum_{k=0}^{n} c_k x^k$, $n = 0, 1, \dots$, then*

$$q_n(x) = \sum_{k=0}^{n} \overline{c}_k s_k(x), \quad n = 0, 1, \dots, \tag{9.9}$$

with

$$\overline{c}_k = \sum_{j=k}^{n} c_j \widehat{s}_{j,k}. \tag{9.10}$$

Proof. Relation (9.9) follows from Proposition 9.1. □

Proposition 9.3 ([103]). *The set G of S. p. s. with umbral composition is a group:*
- *the identity is the sequence $\{x^n\}_n$;*
- *the inverse is the conjugate sequence.*

This group is isomorphic to the exponential Riordan matrix group.

Now, in analogy to the binomial and Appell cases, we consider the conjugate Sheffer vectors

$$S(x) = [s_0(x), \dots, s_n(x), \dots]^T, \quad \widehat{S}(x) = [\widehat{s}_0(x), \dots, \widehat{s}_n(x), \dots]^T, \tag{9.11}$$

and, for $n \in \mathbb{N}$,

$$S_n(x) = [s_0(x), \dots, s_n(x)]^T, \quad \widehat{S}_n(x) = [\widehat{s}_0(x), \dots, \widehat{s}_n(x)]^T. \tag{9.12}$$

Then we have

Proposition 9.4. *With the notations introduced in the previous chapters, the following statements hold:*

$$S(x) = SX(x), \quad \widehat{S}(x) = \widehat{S}X(x); \tag{9.13}$$

$$S(x) = (A \cdot P)X(x) = A \cdot \overline{P}(x), \quad \widehat{S}(x) = (\widehat{P} \cdot \widehat{A})X(x) = (\widehat{A}^* \cdot \widehat{P})X(x) = \widehat{A}^*\overline{\widehat{P}}(x); \tag{9.14}$$

$$X(x) = \widehat{S}S(x), \quad X(x) = S\widehat{S}(x); \tag{9.15}$$

$$\overline{P}(x) = \widehat{A} \cdot S(x), \quad \overline{\widehat{P}}(x) = A^* \cdot \widehat{S}(x); \tag{9.16}$$

where $\widehat{A}^* := (\widehat{a}^*_{i,j})$ is the Appell matrix associated with the sequence $(\widehat{s}_i)$ defined in (2.18), and A^* is its conjugate matrix.

Proof. The proof follows directly from Definition 9.1, Proposition 2.4, and known properties. We note that the decomposition $\widehat{S}(x) = \widehat{A}^*\overline{\widehat{P}}(x)$ easily follows using the definition of the inverse of exponential Riordan matrix (2.24). □

Remark 9.1. We observe that for $n \in \mathbb{N}$, from (9.14)–(9.16), we have:

$$\begin{aligned} S_n(x) &= S_nX_n(x), \quad \widehat{S}_n(x) = \widehat{S}_nX_n(x); \\ X_n(x) &= \widehat{S}_nS_n(x), \quad X_n(x) = S_n\widehat{S}_n(x); \\ S_n(x) &= A_n \cdot \overline{P}_n(x), \quad \widehat{S}_n(x) = \widehat{A}^*_n \cdot \overline{\widehat{P}}_n(x); \\ \overline{P}_n(x) &= \widehat{A}_nS_n(x), \quad \overline{\widehat{P}}_n(x) = A^*_n\widehat{S}_n(x). \end{aligned} \tag{9.17}$$

Hence, from the second line of (9.17), we can get (9.7).

Remark 9.2. We note explicitly that for $a_0 = 1$, $a_i = 0$, $i > 0$, the S. p. s. $\{s_n\}$ is the b. p. s. for b, whereas for $b_1 = 1$, $b_i = 0$, $i > 0$, $\{s_n\}$ is the A. p. s. for a.

Then an important first characterization follows:

Proposition 9.5 ([65]). *The polynomial sequences $\{s_n\}_n$ and $\{\widehat{s}_n\}_n$ are C. S. p. s. if and only if they are the umbral composition of an A. p. s. and a b. p. s., that is, $\forall n \in \mathbb{N}$:*

$$s_n(x) = \sum_{k=0}^{n} a_{n,k}p_k(x) = \sum_{k=0}^{n} \binom{n}{k} a_{n-k}p_k(x) = \sum_{k=0}^{n} \binom{n}{k} a_kp_{n-k}(x) = a_n(p_n(x)), \tag{9.18}$$

$$\widehat{s}_n(x) = \sum_{k=0}^{n} \widehat{a}^*_{n,k}\widehat{p}_k(x) = \sum_{k=0}^{n} \binom{n}{k} \widehat{a}^*_{n-k}\widehat{p}_k(x) = \widehat{a}^*_n(\widehat{p}_n(x)), \tag{9.19}$$

where $\{a_n\}$, $\{\widehat{a}^*_n\}$ and $\{p_n\}$, $\{\widehat{p}_n\}$ are, respectively, A. p. s. and b. p. s.

Moreover, the polynomial sequences $\{p_n\}$ and $\{\widehat{p}_n\}$ are conjugate b. p. s. if and only if they are the umbral composition of an A. p. s. and an S. p. s., that is,

$$p_n(x) = \sum_{j=0}^{n} \binom{n}{j} \widehat{a}_{n-j}s_j(x) = \widehat{a}_n(s_n(x)), \tag{9.20}$$

$$\widehat{p}_n(x) = \sum_{j=0}^{n} \binom{n}{j} a^*_{n-j} \widehat{s}_n(x) = a^*_n(\widehat{s}_n(x)), \tag{9.21}$$

where $\{a_n\}$ *and* $\{a^*_n\}$ *are A. p. s., and* $\{s_n\}$, $\{\widehat{s}_n\}$ *are S. p. s., connected by*

$$a_n(0) = s_n(0), \quad a^*_n(0) = \widehat{s}_n(0).$$

Proof. The result follows observing the latest identities of (9.17). □

Remark 9.3. From Proposition 9.5, it follows that Definition 9.1 is equivalent to Definition given in [65, Remark 3].

Theorem 9.2 (Generating function [65, Theorem 3.6]). *The polynomial sequences* $\{s_n\}_n$, $\{\widehat{s}_n\}_n$ *are C. S. p. s. for* $((a_i), (b_i))_{i\in\mathbb{N}}$ *if and only if the following statements hold:*

$$g(t)\exp(xf(t)) = \sum_{n=0}^{\infty} s_n(x)\frac{t^n}{n!}, \tag{9.22}$$

$$\frac{1}{g(\bar{f}(t))}\exp(x\bar{f}(t)) = \sum_{n=0}^{\infty} \widehat{s}_n(x)\frac{t^n}{n!}, \tag{9.23}$$

where $g(t), f(t), \frac{1}{g(t)}, \bar{f}(t)$ *are defined in* (9.4), (9.5).

Proof. If $\{s_n\}_n$ is an S. p. s. for $((a_i), (b_i))_{i\in\mathbb{N}}$, we have

$$g(t) = \sum_{i=0}^{\infty} a_i \frac{t_i}{i!}, \quad \exp(xf(t)) = \sum_{n=0}^{\infty} p_n(x)\frac{t^n}{n!}.$$

Then multiplying, we get

$$g(t)\exp(xf(t)) = \sum_{n=0}^{\infty}\left(\sum_{i=0}^{n}\binom{n}{i} a_i p_{n-i}(x)\right)\frac{t^n}{n!},$$

and from (9.18), we get the result.

Vice versa, from (9.22), we get the result from Proposition 9.5.

Likewise, we get (9.23). □

The function $g(t)\exp(xf(t))$ is called the *generating function* of S. p. s. $\{s_n\}$, just as $\frac{1}{g(\bar{f}(t))}\exp(x\bar{f}(t))$ is the *generating function* of $\{\widehat{s}_n\}$.

After Theorem 9.2, the C. S. p. s. for $((a_i), (b_i))_{i\in\mathbb{N}}$, $\{s_n\}$ and $\{\widehat{s}_n\}$, given in Definition 9.1, are the polynomial sets of A-type zero, defined by Sheffer in [179, pp. 594–595], [163, p. 218].

9.3 Recurrence relations and determinantal forms

For Sheffer polynomial sequence some recurrence relations and determinantal forms can be derived.

Theorem 9.3 (First recurrence relation and related determinantal form [65]). *The polynomial sequences $\{s_n\}$, $\{\widehat{s}_n\}$ are C. S. p. s. for $((a_i),(b_i))$ if and only if*

$$s_0(x) = s_{0,0}, \quad s_n(x) = \frac{1}{\widehat{s}_{n,n}}\left(x^n - \sum_{k=0}^{n-1} \widehat{s}_{n,k} s_k(x)\right), \tag{9.24}$$

$$\widehat{s}_0(x) = \widehat{s}_{0,0}, \quad \widehat{s}_n(x) = \frac{1}{s_{n,n}}\left(x^n - \sum_{k=0}^{n-1} s_{n,k} \widehat{s}_k(x)\right). \tag{9.25}$$

Moreover, it holds that

$$s_0(x) = \frac{1}{\widehat{s}_{0,0}},$$
$$s_n(x) = \frac{(-1)^n}{\prod_{i=0}^n \widehat{s}_{i,i}} \begin{vmatrix} 1 & x & \dots & x^{n-1} & x^n \\ \widehat{s}_{0,0} & \widehat{s}_{1,0} & \dots & \widehat{s}_{n-1,0} & \widehat{s}_{n,0} \\ & \ddots & & & \vdots \\ & & \ddots & & \vdots \\ & & & \widehat{s}_{n-1,n-1} & \widehat{s}_{n,n-1} \end{vmatrix}, \tag{9.26}$$

$$\widehat{s}_0(x) = \frac{1}{s_{n,n}},$$
$$\widehat{s}_n(x) = \frac{(-1)^n}{\prod_{i=0}^n s_{i,i}} \begin{vmatrix} 1 & x & \dots & x^{n-1} & x^n \\ s_{0,0} & s_{1,0} & \dots & s_{n-1,0} & s_{n,0} \\ & \ddots & & & \vdots \\ & & \ddots & & \vdots \\ & & & s_{n-1,n-1} & s_{n,n-1} \end{vmatrix}. \tag{9.27}$$

Proof. The proof is analogous to the proof of Theorems 3.4, 3.7, 6.3, 6.5. In particular, (9.24) and (9.25) follow from (9.7), (9.8) or equivalently (9.13). Expanding determinant forms respect to the last column, we get the equivalence with (9.24) and (9.25). □

Theorem 9.4 (Second recurrence relation). *Let $\{s_n\}$ be a polynomial sequence. It is an S. p. s. if and only if there exists two numerical sequences (a_i) with $a_0 \neq 0$ and (b_i), with $b_0 = 0$ and $b_1 \neq 0$ such that, if we consider the formal power series $g(t) = \sum_{i=0}^{\infty} a_i \frac{t^i}{i!}$ and $f(t) = \sum_{i=0}^{\infty} b_i \frac{t^i}{i!}$, the following recurrence relation holds:*

$$\begin{aligned}(n+1)\alpha_0 s_{n+1}(x) = {} & (x + \beta_0 - n\alpha_1)s_n(x) + \binom{n}{1}[\beta_1 - (n-1)\alpha_2]s_{n-1}(x) \\ & + \binom{n}{2}[\beta_2 - (n-2)\alpha_3]s_{n-2}(x) + \cdots + \beta_n s_0(x) \quad \forall n \geq 2, \end{aligned} \tag{9.28}$$

where (α_i) and (β_i) are defined by

$$\frac{1}{f'(t)} = \sum_{i=0}^{\infty} \alpha_i \frac{t^i}{i!}, \quad \frac{g'(t)}{g(t)f'(t)} = \sum_{i=0}^{\infty} \beta_i \frac{t^i}{i!}. \tag{9.29}$$

Proof. Let $\{s_n\}$ be the S. p. s. for $((a_i),(b_i))_{i\in\mathbb{N}}$. Then we have the generating function

$$g(t)\exp(xf(t)) = \sum_{n=0}^{\infty} s_n(x)\frac{t^n}{n!}.$$

By partial derivation with respect to the variable t, we get

$$\left[\frac{g'(t)}{g(t)f'(t)} + x\right] f'(t)g(t)\exp(xf(t)) = \sum_{n=0}^{\infty} s_{n+1}(x)\frac{t^n}{n!},$$

and so

$$\left[\frac{g'(t)}{g(t)f'(t)} + x\right] g(t)\exp(xf(t)) = \sum_{n=0}^{\infty} s_{n+1}(x)\frac{t^n}{n!}\cdot\frac{1}{f'(t)}.$$

Then by setting

$$\frac{g'(t)}{g(t)f'(t)} := \sum_{i=0}^{\infty}\beta_i\frac{t^i}{i!}, \quad \frac{1}{f'(t)} := \sum_{i=0}^{\infty}\alpha_i\frac{t^i}{i!},$$

after easy calculation, we have the result.

Vice versa, we observe that (9.29) are two differential equations, easily solvable in a variety of cases [166, p. 158]. Hence we get (a_i), (b_i) and related S. p. s. $\{s_n\}$. □

Remark 9.4. We note that (9.28) also holds for $n = 0, 1$ if we set $s_{-i}(x) \equiv 0$, $i = 1, 2, \ldots, n$.

Remark 9.5. We note that the necessary condition of the above theorem is also in [215], but it is proved with very different methods, and it is independent from (9.22).

Remark 9.6. Given

$$f'(t)\cdot\frac{1}{f'(t)} = 1,$$

we have that the coefficients α_i are given by

$$\sum_{k=0}^{n}\binom{n}{k}\alpha_k b_{n-k+1} = \delta_{n,0}, \quad n = 0, 1, \ldots.$$

The coefficients of the power series

$$\frac{g'(t)}{g(t)f'(t)} := \sum_{i=0}^{\infty}\beta_i\frac{t^i}{i!}$$

are

$$\beta_n = \sum_{k=0}^{n}\binom{n}{k}\alpha_k\bar{\hat{a}}_{n-k},$$

where

$$\bar{\hat{a}}_n = \sum_{k=0}^{n}\binom{n}{k}\hat{a}_k a_{n-k+1}.$$

Remark 9.7. For $g(t) = 1$, (9.28) becomes the second recurrence relation for b. p. s. (3.25), whereas for $f(t) = t$, it is the second recurrence relation for A. p. s. (6.34).

Remark 9.8. If the recurrence relation (9.28) becomes a three-term recurrence, that is,

$$\sum_{k=2}^{n} \binom{n}{k} [\beta_k - (n-k)\alpha_{k+1}] s_{n-k}(x) = 0, \quad n > 2, \quad \forall\, x \in \mathbb{K},$$

then the sequence $\{s_n\}$, on suitable hypothesis, is also orthogonal [90, 92], [40, p. 82].

It is known [92] that the unique S. p. s., which are also orthogonal, are:

- *Laguerre polynomials* $\{L_n^{(\alpha)}\}$

$$\sum_{n=0}^{\infty} L_n^{(\alpha)}(x) \frac{t^n}{n!} = (1-t)^{-(\alpha+1)} \exp\left(\frac{xt}{t-1}\right), \quad \alpha > -1,$$

- *Hermite polynomials* $\{H_n\}$

$$\sum_{n=0}^{\infty} H_n(x) \frac{t^n}{n!} = \exp\left(xt - \frac{t^2}{4}\right),$$

- *Charlier polynomials* $\{C_n^{(\alpha)}\}$

$$\sum_{n=0}^{\infty} C_n^{(\alpha)}(x) \frac{t^n}{n!} = \exp(t)\left(1 + \frac{t}{\alpha}\right)^x,$$

- *Meixener polynomials* $\{M_n(\beta, c)\}$

$$\sum_{n=0}^{\infty} (\beta)_n M_n(\beta, c, x) \frac{t^n}{n!} = \left(1 - \frac{t}{c}\right)^x (1-t)^{-(x+\beta)},$$

- *Meixener–Pollaczek polynomials* $\{P_n^{(\lambda)}(\phi)\}$

$$\sum_{n=0}^{\infty} P_n^{(\lambda)}(\phi, x) \frac{t^n}{n!} = (1 - \exp(i\phi)t)^{-\lambda+ix} (1 - \exp(-i\phi)t)^{-\lambda-ix},$$

- *Krawtchouk polynomials* $\{K_n(p, N)\}$

$$\sum_{n=0}^{\infty} \binom{N}{n} K_n(p, N, x) \frac{t^n}{n!} = \left(1 - \frac{1-p}{p} t\right)(1+t)^{N-x}.$$

Theorem 9.5 (Second determinantal form). *Let $\{s_n\}_n$ be a polynomial sequence. It is an S. p. s. if and only if there exist numerical sequences $(a_i)_i$ with $a_0 \neq 0$, and $(b_i)_i$ with*

$b_0 = 0$, $b_1 \neq 0$, *such that if* $(\beta_i)_i$ *and* (α_i) *are defined as in (9.29), then*

$$
\begin{aligned}
s_0(x) &= 1,\\
s_{n+1}(x) &= \frac{(-1)^{n+1}}{(\alpha_0)^{n+1}(n+1)!}\\
&\times \begin{vmatrix}
x+\beta_0 & -\alpha_0 & \cdots & 0 & 0\\
\beta_1 & x+\beta_0-\alpha_1 & \cdots & 0 & 0\\
\beta_2 & \binom{2}{1}[\beta_1-\alpha_2] & \cdots & 0 & 0\\
\vdots & & \ddots & \ddots & \vdots\\
\beta_{n-1} & \binom{n-1}{1}[\beta_{n-2}-\alpha_{n-1}] & \cdots & \binom{n-1}{n-1}[\beta_0-(n-1)\alpha_1] & -\alpha_0\\
\beta_n & \binom{n}{1}[\beta_{n-1}-\alpha_n] & \cdots & \binom{n}{n-1}[\beta_0-(n-1)\alpha_1] & x+\beta_0-n\alpha_1
\end{vmatrix}.
\end{aligned}
\tag{9.30}
$$

Proof. The proof follows using Theorem 9.4 by applying Cramer's rule for the solution of the first $n+1$ equations in the variables $s_1(x), s_2(x), \dots, s_{n+1}(x)$ of the infinite linear system given by (9.28).

Vice versa, from (9.30), we get (9.28), with Laplace expansion of the determinant with respect to the last row. □

Remark 9.9. It is possible to verify that the matrix $\widehat{R} := (\widehat{r}_{i,j})$, defined as

$$\widehat{r}_{i,j} = \binom{i}{j}[\beta_{i-j} - j\alpha_{i-j+1}],$$

where $(\alpha)_i$ and $(\beta_i)_i$ are as in (9.29), is the production matrix of $\widehat{S}$, the conjugate of S, as given in Definition 9.1.

Remark 9.10. Also the second determinantal form for an S. p. s., as a necessary condition, is in [213, 215], but it is determined by a more general and complex procedure.

A third recurrence relation and the related determinantal form can be determined by expliciting an S. p. s. in the basis of b. p. s., that is:

Theorem 9.6 (Third recurrence relation and related determinantal form, [65]). *The polynomial sequences* $\{s_n\}$, $\{\widehat{s}_n\}$ *are C. S. p. s. for* $((a_i),(b_i))$ *if and only if*

$$s_0(x) = \frac{1}{a_0(x)}, \quad s_n(x) = \frac{1}{\widehat{a}_{n,n}}\left(p_n(x) - \sum_{k=0}^{n-1}\binom{n}{k}\widehat{a}_{n-k}s_k(x)\right), \tag{9.31}$$

$$\widehat{s}_0(x) = \frac{1}{\widehat{a}_0^*(x)}, \quad \widehat{s}_n(x) = \frac{1}{a_{n,n}^*}\left(\widehat{p}_n(x) - \sum_{k=0}^{n-1}\binom{n}{k}a_{n-k}^*\widehat{s}_k(x)\right), \tag{9.32}$$

with $\{p_n\}$, $\{\widehat{p}_n\}$ *conjugate b. p. s. for* $(b_i)_i$, *whereas* $(\widehat{a}_{n,k})$ *and* $(a_{n,k}^*)$ *are defined in Proposition 9.4.*

Moreover, we have

$$s_0(x) = \frac{1}{a_0(x)}, \quad s_n(x) = \frac{(-1)^{n+1}}{\widehat{a}_0^{n+1}} \begin{vmatrix} p_0(x) & p_1(x) & \cdots & p_{n-1}(x) & p_n(x) \\ \widehat{a}_0 & \widehat{a}_1 & \cdots & \widehat{a}_{n-1} & \widehat{a}_n \\ & \ddots & & & \vdots \\ & & \ddots & & \vdots \\ & & & \widehat{a}_0 & \binom{n}{n-1}\widehat{a}_1 \end{vmatrix}, \tag{9.33}$$

$$\widehat{s}_0(x) = \frac{1}{\widehat{a}_0^*(x)}, \quad \widehat{s}_n(x) = \frac{(-1)^{n+1}}{(a_0^*)^{n+1}} \begin{vmatrix} \widehat{p}_0(x) & \widehat{p}_1(x) & \cdots & \widehat{p}_{n-1}(x) & \widehat{p}_n(x) \\ a_0^* & a_1^* & \cdots & a_{n-1}^* & a_n^* \\ & \ddots & & & \vdots \\ & & \ddots & & \vdots \\ & & & a_0^* & \binom{n}{n-1}a_1^* \end{vmatrix}. \tag{9.34}$$

Proof. For the proof, we recall (9.17). In fact, we get

$$\overline{P}_n(x) = \widehat{A}_n S_n(x), \quad \overline{\widehat{P}}_n(x) = A_n^* \widehat{S}_n(x),$$

from which the result follows after easy calculation.

Moreover, the determinantal forms (9.33) and (9.34) follow from (9.31) and (9.32), with the same technique of Theorems 9.3. □

Remark 9.11. We note that by first and third determinantal forms, an S. p. s. is a determinant of a particular Hessenberg matrix, whereas for the second determinantal form, it is the characteristic polynomial of a suitable Hessenberg matrix.

9.4 Differential equations

The generating function (9.22) and the recurrence relations of an S. p. s. allow us to give some differential equations.

Proposition 9.6 (Derivation). *Let $\{s_n\}_n$ be the S. p. s. for $((a_i),(b_i))_{i\in\mathbb{N}}$. Then we get*

$$\frac{1}{k!}s_n^{(k)}(x) = \sum_{j=k}^{n} \binom{n}{j} p_{j,k} s_{n-j}(x), \quad k = 0,\ldots,n, \tag{9.35}$$

where $p_{j,k}$ are defined in (3.1).

Proof. From the generating function of $\{s_n\}$, after k derivations with respect to the variable x, we get

$$(f(t))^k g(t)\exp(xf(t)) = \sum_{k=0}^{\infty} s_n^{(k)}(x)\frac{t^n}{n!}. \tag{9.36}$$

From (2.3), we have

$$(f(t))^k = k!\sum_{n=k}^{\infty} p_{n,k}\frac{t^n}{n!}, \quad k = 0,\dots,n.$$

Then by substituting it into (9.36) and applying the Cauchy product, we have the result. □

Remark 9.12. We note explicitly that for $k = 1$, we have

$$s_n'(x) = \sum_{j=1}^{n}\binom{n}{j} b_j s_{n-j}(x) = \sum_{j=0}^{n-1}\binom{n}{j} b_{n-j} s_j(x), \quad s_n(0) = a_n, \quad n = 0,1,\dots. \tag{9.37}$$

Remark 9.13. For $b_1 = 1$ and $b_i = 0$ for $i > 1$, the S. p. s. $\{s_n\}$ is an A. p. s. [17], and (9.37) is the known formula (6.9). Likewise,, for $a_0 = 1$ and $a_i = 0$ for $i > 0$, the S. p. s. $\{s_n\}$ is a b. p. s., and (9.37) is formula (3.3).

Corollary 9.1. *For a S. p. s. $\{s_n\}$, we get the following recurrence relation for the coefficients:*

$$s_{n,k+1} = \frac{1}{k+1}\sum_{i=k}^{n}\binom{n}{i} b_{n-i} s_{i,k} = \frac{1}{k+1}\sum_{i=1}^{n-k}\binom{n}{i} b_i s_{n-i,k}, \quad k = 0,\dots,n-1, \tag{9.38}$$

with

$$\begin{aligned} s_{n,0} &= a_n, \\ s_{n,n} &= a_0 b_1^n, \end{aligned} \quad n = 0,1,\dots. \tag{9.39}$$

Proof. The proof follows after easy calculation by (9.37). □

Remark 9.14. We note that (9.38) and (9.39) are equivalent to (2.16). That is, the entries of Sheffer-type matrix can be defined by recurrence (9.38)–(9.39).

Remark 9.15. We also note that b. p. s. and S. p. s. associated have the same recurrence relation for the coefficients in the decomposition on basic monomials, but different initial conditions.

As a consequence of the corollary, we can enunciate the following new characterization:

Theorem 9.7. *A polynomial sequence $\{s_n\}$ is an S. p. s. if and only if there exist two numerical sequences (a_i), $a_0 \neq 0$, and (b_i), $b_0 = 0$, $b_1 \neq 0$, such that*

$$\begin{cases} s_0(x) = a_0 \neq 0, & \\ s_n'(x) = \sum_{j=1}^{n}\binom{n}{j} b_j s_{n-j}(x), & n \geq 1, \\ s_n(0) = a_n, & n \geq 1. \end{cases}$$

If $s_n(x) = \sum_{k=0}^{n} s_{n,k}x^k$, *we have*

$$\begin{cases} s_n(0) = s_{n,0} = a_n, & n = 0,1,\ldots, \\ s_{n,n} = a_0 b_1^n, & \\ s_{n,k+1} = \frac{1}{k+1}\sum_{i=k}^{n}\binom{n}{i} b_{n-i}s_{i,k} = \frac{1}{k+1}\sum_{i=1}^{n-k}\binom{n}{i} b_i s_{n-i,k}, & k = 0,\ldots,n-2. \end{cases}$$

Proof. The proof follows from previous results. □

Proposition 9.7. *Let* $\{s_n\}_n$ *be the S. p. s. for* $((a_i)_i, (b_i)_i)$. *If* $s_n(x) \geq 0$, $\forall x \geq 0$, *then* $a_i \geq 0$, $i = 1,2,\ldots$, $a_0 > 0$, $b_1 > 0$.

Proof. The result follows from (9.39). □

Theorem 9.7 establishes the connection between Appell polynomials (defined by Appell [17] in 1880), the successive Sheffer polynomials [179] (1939), and the binomial polynomials in Definition 3.1.

Now we consider the following vectors:

$$\overline{S}_n(x) = \left[\binom{n}{0} s_n(x), \ldots, \binom{n}{j} s_{n-j}(x), \ldots \binom{n}{n} s_0(x)\right]^T, \tag{9.40}$$

$$D\overline{S}_n(x) = \left[\frac{s_n^{(0)}(x)}{0!}, \ldots, \frac{s_n^{(j)}(x)}{j!}, \ldots, \frac{s_n^{(n)}(x)}{n!}\right]^T. \tag{9.41}$$

Proposition 9.8. *With the previous notations, the following statements hold:*

$$D\overline{S}_n(x) = P^T \overline{S}_n(x), \tag{9.42}$$

$$\overline{S}_n(x) = \widehat{P}^T D\overline{S}_n(x). \tag{9.43}$$

Proof. Identity (9.42) follows from Proposition 9.6, according to (9.40) and (9.41). Moreover, P^T is nonsingular, and thus

$$(P^T)^{-1} = (P^{-1})^T = \widehat{P}^T,$$

and hence relation (9.43) follows. □

Now we can prove that (9.35) is invertible, as follows:

Proposition 9.9 (Fourth recurrence relation). *For S. p. s.* $\{s_n\}$ *with the previous notations, we have*

$$\binom{n}{i} s_{n-i}(x) = \sum_{j=i}^{n} \widehat{p}_{j,i}\frac{s_n^{(j)}(x)}{j!}, \quad n = 0,1,\ldots, \quad i = 0,\ldots,n, \tag{9.44}$$

where $\widehat{p}_{i,j}$ *are defined in Algorithm 2.1.2.*

Proof. Relation (9.44) follows from (9.43). □

Remark 9.16. We note that (9.44) can be rewritten as

$$\binom{n}{i} s_i(x) = \sum_{j=n-i}^{n} \widehat{p}_{j,n-i} \frac{s_n^{(j)}(x)}{j!}, \qquad n = 0, 1, \dots, \qquad i = 0, \dots, n. \tag{9.45}$$

Proposition 9.10 (Sheffer integration). *For $n \in \mathbb{N}$ and $i = 0, \dots, n$, the following statement holds:*

$$\begin{aligned}\int_0^x s_i(t)dt &= \frac{i!(n-i)!}{n!} \sum_{j=n-i}^{n} \widehat{p}_{j,n-i} \sum_{m=j-1}^{n} \binom{n}{m} p_{n-i,m} \left[\frac{s_{n-j}(x) - s_{n-j}(0)}{j} \right] \\ &= \sum_{k=0}^{i} s_{ik} \frac{x^{k+1}}{k+1}. \end{aligned}\tag{9.46}$$

Proof. The proof follows after easy calculation from (9.44) and from Definition 9.1. □

Relation (9.45), combined with recurrence relations, allows us to write linear differential equations for Sheffer sequences. In fact, the following holds:

Theorem 9.8 (First differential equation). *Let $\{s_n\}$ be the S. p. s. for $((a_i), (b_i))_{i \in \mathbb{N}}$. Then $\forall n \in \mathbb{N}$, $s_n(x)$ satisfies the linear differential equation of order n:*

$$\widehat{s}_{n,n} y(x) + \widehat{s}_{n,n} \sum_{k=1}^{n} c_{n,k} \frac{y^{(k)}(x)}{k!} = x^n, \tag{9.47}$$

where

$$c_{n,k} := \sum_{j=n-k}^{n} \frac{\widehat{s}_{n,n-j}}{\binom{n}{j}} \widehat{p}_{k,j}, \qquad k = 0, \dots, n. \tag{9.48}$$

Proof. The proof follows, after easy calculation, by substituting (9.45) into the first recurrence relation. □

Theorem 9.9 (Second differential equation). *Let $\{s_n\}$ be the S. p. s. for $((a_i), (b_i))_{i \in \mathbb{N}}$. Then $\forall n \in \mathbb{N}$, $s_n(x)$ satisfies the linear differential equation of order n:*

$$n\alpha_0 y(x) - \frac{1}{n}(x + \beta_0 - (n-1)\alpha_1)\widehat{b}_1 y'(x) - \sum_{j=2}^{n} \left[\frac{1}{n}(x + \beta_0 - (n-1)\alpha_1)\widehat{b}_j + c_{n,j} \right] \frac{y^{(j)}(x)}{j!} = 0, \tag{9.49}$$

where (α_i) and (β_i) are defined as in (9.29), and

$$c_{n,j} := \sum_{k=2}^{j} \frac{k-1}{n} [\beta_{k-1} - (n-k)\alpha_k] \widehat{b}_{j,k}. \tag{9.50}$$

Proof. The proof follows, after easy calculation, by substituting (9.45) in the second recurrence relation. □

We define the matrix $\mathcal{SD} := (d_{i,j})$, $i = 0, 1 \ldots, j = 0, \ldots, i$, with

$$d_{i,j} = \binom{i}{j} b_{i-j}$$

and the vector

$$S'_n(x) = [s'_0(x), \ldots, s'_n(x)]^T.$$

Then we get the following:

Proposition 9.11 (Differentiation matrix). *With previous notations, we have*

$$S'_n(x) = (\mathcal{SD})S_n(x). \tag{9.51}$$

Proof. The proof follows from (9.37). □

The matrix $\mathcal{SD}$ is called the *Sheffer differentiation matrix*; it coincides with the differentiation matrix for the b. p. s. associated. Moreover, for $b_1 = 1$ and $b_i = 0$, $i = 2, \ldots$, (9.51) is the Appell differentiation matrix.

9.5 General properties

With elementary tools of matrix calculus, we can prove some general properties of Sheffer sequences.

Theorem 9.10 (Sheffer identity [65]). *The polynomial sequences $\{s_n\}$ and $\{\widehat{s}_n\}$ are C. S. p. s. for $((a_i), (b_i))_{i\in\mathbb{N}}$ if and only if the following identities hold:*

$$s_n(x+y) = \sum_{i=0}^{n} \binom{n}{i} s_{n-i}(x)p_i(y) = \sum_{i=0}^{n} \binom{n}{i} s_i(x)p_{n-i}(y) \quad \forall n \in \mathbb{N}, \forall x, y \in \mathbb{R}, \tag{9.52}$$

$$\widehat{s}_n(x+y) = \sum_{i=0}^{n} \binom{n}{i} \widehat{s}_{n-i}(x)\widehat{p}_i(y) = \sum_{i=0}^{n} \binom{n}{i} \widehat{s}_i(x)\widehat{p}_{n-i}(y) \quad \forall n \in \mathbb{N}, \forall x, y \in \mathbb{R}, \tag{9.53}$$

where $\{p_n\}$ and $\{\widehat{p}_n\}$ are the conjugate b. p. s. for b and $\widehat{b}$.

Proof. The proof follows from determinantal forms (9.33) and, respectively, (9.34), by applying the linearity and the binomial identity of b. p. s. □

Remark 9.17. From Theorem 9.10, the Definition 9.1 is equivalent to the definition given in [167, p. 139].

Corollary 9.2. *The sequences $\{s_n\}$ and $\{\widehat{s}_n\}$ are C. S. p. s. for $((a_i), (b_i))_{i\in\mathbb{N}}$ if and only if*

$$s_n(x) = \sum_{k=0}^{n} \binom{n}{k} s_{n-k}(0)p_k(x), \quad \widehat{s}_n(x) = \sum_{k=0}^{n} \binom{n}{k} \widehat{s}_{n-k}(0)\widehat{p}_k(x). \tag{9.54}$$

Proof. The result follows from Theorem 9.10 by interchanging x and y and by setting $y = 0$. □

We note that (9.54) coincides with (9.18).

Theorem 9.11 (Multiplication theorem, [65, Theorem 3.5]). *A polynomial sequence $\{s_n\}$ is an S. p. s. for $((a_i), (b_i))_{i\in\mathbb{N}}$ if and only if*

$$s_n(mx) = \sum_{i=0}^{n} \binom{n}{i} s_{n-i}(x) p_i((m-1)x), \quad n = 0, 1, \ldots, \quad m = 1, 2, \ldots. \tag{9.55}$$

Proof. The result follows from (9.52) for $y = (m-1)x$. □

9.6 Relationship with δ-operators and functionals

In order to establish a connection with Roman theory, let $\{s_n\}$ and $\{\widehat{s}_n\}$ be C. S. p. s. For (9.2), we can consider the δ-operators Q and $\widehat{Q}$ with properties (9.3).

Then

Theorem 9.12 ([65, Theorem 4.1]). *With the above notations the following statements hold:*

$$Qs_n(x) = ns_{n-1}(x), \quad \widehat{Q}\widehat{s}_n(x) = n\widehat{s}_{n-1}(x), \quad n = 1, 2, \ldots. \tag{9.56}$$

Proof. Using the property of linearity, we can apply the operator Q to the determinantal forms (9.33) or (9.34). We expand the resulting determinant with respect to the first column, and we identify the factor $s_{n-1}(x)$ after multiplication of the ith row by $i-1$, $i = 1, \ldots, n$, and the jth column by $\frac{1}{j}$, $j = 1, \ldots, n$. □

Now we consider new operators.

Definition 9.2. Let T and $\widehat{T}$ be the operators defined by

$$\forall n \in \mathbb{N}, \quad Tx^n = \sum_{i=0}^{n} \binom{n}{i} \widehat{a}_i x^{n-i} = \widehat{a}_n(x), \quad \widehat{T}x^n = \sum_{i=0}^{n} \binom{n}{i} a_i x^{n-i} = a_n(x), \tag{9.57}$$

where $(a_i)_i$ and $(\widehat{a}_i)_i$ are the numerical sequences that generate the A. p. s. $\{a_n\}$ and $\{\widehat{a}_n\}$.

Then we have:

Theorem 9.13. *Let $\{s_n\}$ and $\{\widehat{s}_n\}$ be C. S. p. s. for $((a_i), (b_i))_{i\in\mathbb{N}}$. The following relations hold:*

$$T(s_n)(x) = p_n(x), \quad \widehat{T}(\widehat{s}_n)(x) = \widehat{p}_n(x), \tag{9.58}$$

where T and $\widehat{T}$ are the operators defined in (9.57), and $\{p_n\}$ and $\{\widehat{p}_n\}$ are the b. p. s. associated respectively with (b_i) and $(\widehat{b}_i)_i$.

Proof. The result follows by standard calculations and using Proposition 9.5 or Remark 9.1. □

Remark 9.18. We note that the operators T and $\widehat{T}$ as defined in (9.57) associate with the monomial x^n the conjugate Appell polynomials $\widehat{a}_n(x)$ and $a_n(x)$. Then the Theorem 9.13 is equivalent, in operator language, to Proposition 9.5.

The differential equation (9.56) allows us to write the following functional equation:

Theorem 9.14 (Functional equation [65, Theorem 4.4]). *Let $\{s_n\}$ and $\{\widehat{s}_n\}$ be C. S. p. s., and let Q and $\widehat{Q}$ be the δ-operators with associated b. p. s. $\{p_n\}$ and $\{\widehat{p}_n\}$. Then $s_n(x)$ and $\widehat{s}_n(x)$ satisfy the functional equations*

$$\frac{\widehat{a}_n}{n!}Q^n y(x) + \frac{\widehat{a}_{n-1}}{(n-1)!}Q^{n-1}y(x) + \cdots + Q_0 y(x) = p_n(x), \tag{9.59}$$

$$\frac{a_n}{n!}\widehat{Q}^n y(x) + \frac{a_{n-1}}{(n-1)!}\widehat{Q}^{n-1}y(x) + \cdots + \widehat{Q}_0 y(x) = \widehat{p}_n(x). \tag{9.60}$$

Proof. Relation (9.59) follows by combining (9.56) and (9.24), just as for (9.60). □

Now we can consider:

1. the linear functionals L, $\widehat{L}$ on $\mathcal{P}_n$ defined by

$$\forall n \in \mathbb{N} \quad : \quad L(x^i) = \widehat{a}_i, \quad \widehat{L}(x^i) = a_i, \quad i = 0, \ldots, n, \tag{9.61}$$

   with $a_0 \neq 0$ and $\widehat{a}_0 \neq 0$.
   We note that they are invertible functionals.
2. the linear functionals M, $\widehat{M}$ on $\mathcal{P}_n$ defined by

$$\forall n \in \mathbb{N} \quad : \quad M(x^i) = \widehat{b}_i, \quad \widehat{M}(x^i) = b_i, \quad i = 0, \ldots, n, \tag{9.62}$$

   with $b_0 = 0$, $b_1 \neq 0$ and $\widehat{b}_0 = 0$, $\widehat{b}_1 \neq 0$.
   They are δ-functionals.

Then for the powers of linear functional, defined in (3.37) and (3.38), we have

Theorem 9.15. *With previous notation, $\{s_n\}$, $\{\widehat{s}_n\}$ are C. S. p. s. for $((a_i), (b_i))_i$ if and only if*

$$\forall n \in \mathbb{N}, \quad LM^k(s_n(x)) = n!\delta_{n,k}, \quad k = 0, \ldots, n; \tag{9.63}$$

$$\widehat{L}\widehat{M}^k(\widehat{s}_n(x)) = n!\delta_{n,k}, \quad k = 0, \ldots, n. \tag{9.64}$$

Proof. Observing that, according to (3.37)–(3.40),

$$LM^k(x^i) = k!\widehat{S}_{i,k}, \tag{9.65}$$

statement (9.63) follows from (9.26). Likewise, we get (9.64). □

After Theorem 9.15, we also say that the polynomial sequences $\{s_n\}$, $\{\widehat{s}_n\}$ are C. S. p. s. for the linear functionals L, M.

We set

$$\forall n \in \mathbb{N}, \quad k = 0, \dots, n, \quad T_k = LM^k, \quad \widehat{T}_k = \widehat{L}\widehat{M}^k. \tag{9.66}$$

They are linear independent functionals on $\mathcal{P}_n$. Then we have the following:

Theorem 9.16. *The C. S. p. s. $\{s_n\}$, $\{\widehat{s}_n\}$ are the unique solution, respectively, of general linear interpolant problems on $\mathcal{P}_n$:*

$$\forall n \in \mathbb{N}, \quad T_k(q_n) = n!\delta_{n,k}, \quad \widehat{T}_k(q_n) = n!\delta_{n,k}, \quad k = 0, \dots, n. \tag{9.67}$$

Proof. The result follows using a theorem on general linear interpolant [76, p. 35] and from the first determinantal forms. □

Corollary 9.3. *The coefficients $s_{n,i}$ of the polynomial $s_n(x)$ satisfy the linear system*

$$\sum_{i=0}^{n} s_{n,i}\widehat{s}_{i,k} = n!\delta_{n,k}, \quad k = 0, \dots, n. \tag{9.68}$$

Proof. It follows from (9.63) and (9.64). □

From Theorem 9.16 and Corollary 9.3, the S. p. s. defined by Roman [166, p. 17] coincides with the Definition 9.1.

The following theorem characterizes the generating function by the linear functionals L, $\widehat{L}$, M, $\widehat{M}$.

Theorem 9.17 (Generating function). *The polynomial sequences $\{s_n\}$, $\{\widehat{s}_n\}$ are C. S. p. s. for the linear functionals L, M if and only if*

$$\frac{\sum_{k=0}^{\infty} x^k M_x^k(\exp(xt))}{L_x(\exp(xt))} = g(t)\exp(xf(t)) = \sum_{n=0}^{\infty} s_n(x)\frac{t^n}{n!}; \tag{9.69}$$

$$\frac{\sum_{k=0}^{\infty} x^k \widehat{M}_x^k(\exp(xt))}{\widehat{L}_x(\exp(xt))} = \frac{1}{g(\bar{f}(t))}\exp(x\bar{f}(t)) = \sum_{n=0}^{\infty} \widehat{s}_n(x)\frac{t^n}{n!}. \tag{9.70}$$

Proof. The proof follows, after calculations, from (9.61) and (9.62). □

9.7 Summary

The Sheffer polynomial sequence has been introduced by particular infinite, lower triangular matrix, which results from binomial-type and Appell-type matrices. Then more characterizations are given, some of which are known. Recurrence, differential, and determinantal forms are considered, just as previous polynomial sequences. Finally, a link to the theory of Sheffer and Rota et al. is given.

10 Applications to linear interpolation and operators approximation theory

Abstract: We consider the extension of Appell interpolation problem to more general Sheffer polynomials. Moreover, the approximation theory for Százs-type operators with Sheffer sequences is also considered.

10.1 Linear interpolation

In Chapter 8, we considered the Appell interpolation. Now we will extend this approach to more general Sheffer sequences, following [64]. For this purpose, let X be the linear space of real continuous functions defined in the interval $[a, b]$ with continuous derivatives of all necessary orders.

Let $\mathcal{P}_n \subset X$, $n \in \mathbb{N}$, be the space of polynomials of degree less than or equal to n. Let Q be a linear operator on X of the form

$$Qf = \sum_{i=1}^{\infty} c_i \frac{f^{(i)}}{i!}, \quad c_1 \neq 0. \tag{10.1}$$

This operator on $\mathcal{P}_n$ is a δ-operator [167, p. 129].

Defining

$$Q^0 f = If, \quad Q^1 f = Qf, \quad Q^i f = Q(Q^{i-1}(f)), \quad i = 2, 3, \dots, \tag{10.2}$$

we consider the set

$$XQ = \{f \in X \mid Q^i f \in X, \quad i = 0, \dots, n, \quad \forall n \in \mathbb{N}\}. \tag{10.3}$$

Let $\mathcal{L}$ be a linear functional on X with $\mathcal{L}(1) \neq 0$. Then, for each $f \in XQ$, we want to look for a polynomial $P_n[f]$ of degree less than or equal to n such that

$$f = P_n[f] + R_n[f] \tag{10.4}$$

with

$$\mathcal{L}(Q^i f) = \mathcal{L}(Q^i P_n[f]), \quad i = 0, \dots, n. \tag{10.5}$$

We will call this problem umbral or Sheffer interpolant problem [64] because its solution, if it exists, can be expressed by a basis of Sheffer polynomials, also called umbral basis.

For the solution of this problem, we observe that given the operator Q (10.1) and setting $\widehat{b}_i = c_i$, $i = 1, \dots$, $\widehat{b}_0 = 0$, we get the numerical sequence $(b_i)_i$ from (2.9). If

https://doi.org/10.1515/9783110652925-010

$\forall\, i \in \mathbb{N}$, we set $\widehat{s}_i = \mathcal{L}(x^i)$, we then have the numerical sequence $(\widehat{s}_i)$. Then from (2.19), (2.20), (9.26), (9.27), we can consider the C. S. p. s. $\{s_n\}$, $\{\widehat{s}_n\}$. In this case, we call the sequences $\{s_n\}$ and $\{\widehat{s}_n\}$ C. S. p. s. for $(\mathcal{L}, Q)$.

Now we consider $f \in \mathcal{P}_n$, then we get:

Theorem 10.1. *With the previous notation, let $\omega_i \in \mathbb{K}$, $i = 0, \dots, n$. The polynomials*

$$P_n(x) := \sum_{i=0}^{n} \omega_i s_i(x), \quad \overline{P}_n(x) := \sum_{i=0}^{n} \omega_i \widehat{s}_i(x) \tag{10.6}$$

are the unique polynomials of degree less than or equal to n such that for $i = 0, \dots, n$, respectively,

$$\mathcal{L}(Q^i P_n) = i!\omega_i, \quad \mathcal{L}(\widehat{Q}^i \overline{P}_n) = i!\omega_i.$$

Proof. From the hypothesis, applying (9.56) and (9.27), we have

$$\mathcal{L}(Q^i s_n(x)) = n(n-1)\cdots(n-i+1)\mathcal{L}(s_{n-i}(x)) = i!\binom{n}{i}\delta_{n,i}.$$

Hence the result follows from the linearity of Q. In analogy, we get the result for $\widehat{Q}$. □

Corollary 10.1 (Representation theorem). *For each $P_n(x) \in \mathcal{P}_n$, we have*

$$P_n(x) = \sum_{i=0}^{n} \frac{\mathcal{L}(Q^i P_n)}{i!} s_i(x) = \sum_{i=0}^{n} \frac{\mathcal{L}(\widehat{Q}^i P_n)}{i!} \widehat{s}_i(x). \tag{10.7}$$

Proof. The proof follows setting in Theorem 10.1

$$\omega_i = \frac{\mathcal{L}(Q^i P_n)}{i!}, \quad i = 0, \dots, n,$$

or

$$\omega_i = \frac{\mathcal{L}(\widehat{Q}^i P_n)}{i!}, \quad i = 0, \dots, n.$$ □

The polynomial (10.7) is very similar to the classic Taylor polynomial for $P_n(x)$.

Theorem 10.2 (Taylor–Sheffer polynomial). *With the previous notations and hypotheses, for each $P_n \in \mathcal{P}_n$, we have*

$$P_n(x) = \sum_{k=0}^{n} \frac{\mathcal{L}(P_n^{(k)})}{k!} z_k(x), \tag{10.8}$$

where

$$z_k(x) = \sum_{i=0}^{k} k!\widehat{b}_{k,i} s_i(x). \tag{10.9}$$

Proof. Given that $QP_n(x) = \sum_{i=0}^{n} \widehat{b}_i \frac{P_n^{(i)}(x)}{i!}$, from (2.3), we get

$$Q^k P_n(x) = \sum_{i=k}^{n} k! \widehat{b}_{i,k} \frac{P_n^{(i)}(x)}{i!}, \tag{10.10}$$

and by easy calculations, the result follows applying Theorem 10.1 and Corollary 10.1. □

The representation (10.8) is very similar to the classical Taylor polynomial. Therefore, we call it the *Taylor polynomial for* $(\mathcal{L}, Q)$ or the *Taylor–Sheffer polynomial.*

Corollary 10.2. *If* $\mathcal{L}(P_n) = P_n(0)$ *and* $Q = D$, *then the S. p. s.* $\{s_i\}$ *is* $\{x^i\}$, *and polynomial* (10.8) *is the classical Taylor polynomial.*

Let us consider a function $f \in XQ$. Then we have the following:

Theorem 10.3. *The polynomials*

$$P_n[f](x) := \sum_{i=0}^{n} \frac{\mathcal{L}(Q^i f)}{i!} s_i(x), \quad \overline{P}_n[f](x) := \sum_{i=0}^{n} \frac{\mathcal{L}(\widehat{Q}^i f)}{i!} \widehat{s}_i(x) \tag{10.11}$$

are the unique polynomials of degree less than or equal to n such that, respectively,

$$\mathcal{L}(Q^i P_n[f])(x) = \mathcal{L}(Q^i f), \quad \mathcal{L}(\widehat{Q}^i \overline{P}_n[f]) = \mathcal{L}(\widehat{Q}^i f), \quad i = 0, \ldots, n. \tag{10.12}$$

Proof. The result follows by Theorem 10.1. □

Remark 10.1. We observe from hypothesis (10.3) that polynomials (10.11) can be written in the form

$$P_n[f](x) := \sum_{i=0}^{n} \mathcal{L}(g_i) \frac{s_i(x)}{i!}, \quad \overline{P}_n[f](x) := \sum_{i=0}^{n} \mathcal{L}(\overline{g}_i) \frac{\widehat{s}_i(x)}{i!}, \tag{10.13}$$

where

$$g_i(x) = Q^i f = \sum_{k=i}^{\infty} i! \widehat{b}_{ki} \frac{f^{(k)}(x)}{k!}, \quad \overline{g}_i(x) = \overline{Q}^i f = \sum_{k=i}^{\infty} i! b_{ki} \frac{f^{(k)}(x)}{k!}.$$

Then we have the following:

Definition 10.1. The polynomials $P_n[f](x)$, $\overline{P}_n[f](x)$ are called the *umbral interpolant polynomials* of the function f for $(\mathcal{L}, Q)$.

Therefore, an estimation of the remainder is useful:

$$R_n[f](x) = f(x) - P_n[f](x), \quad \overline{R}_n[f](x) = f(x) - \overline{P}_n[f](x), \quad \forall x \in (a, b). \tag{10.14}$$

For this purpose we get:

Theorem 10.4. *For any $f(x) \in \mathcal{P}_n$ and $x \in [a,b]$,*

$$\begin{aligned} R_n[f](x) &= 0, \quad R_n[t^{n+1}](x) \neq 0; \\ \overline{R}_n[f](x) &= 0, \quad \overline{R}_n[t^{n+1}](x) \neq 0. \end{aligned} \tag{10.15}$$

Proof. The result follows from Corollary 10.1. □

For a fixed x, we may consider the remainder $R_n[f]$ or $\overline{R}_n[f]$ as a linear functional, which vanishes for any polynomial of degree n.

Therefore, from Peano's theorem [76, p. 69], we have

Theorem 10.5. *Let $f \in C^{n+1}[a,b]$. The following relations hold:*

$$\begin{aligned} R_n[f](x) &= \frac{1}{n!}\int_0^b K_n(x,t) f^{(n+1)}(t)dt, \\ \overline{R}_n[f](x) &= \frac{1}{n!}\int_0^b \overline{K}_n(x,t) f^{(n+1)}(t)dt, \end{aligned} \tag{10.16}$$

where

$$\begin{aligned} K(x,t) &= R_n[(x-t)_+^n](x) = (x-t)_+^n - \sum_{i=0}^{n} \frac{\mathcal{L}_x(Q^i(x-t)_+^n)}{i!} s_i(x), \\ \overline{K}(x,t) &= \overline{R}_n[(x-t)_+^n](x) = (x-t)_+^n - \sum_{i=0}^{n} \frac{\mathcal{L}_x(\widehat{Q}^i(x-t)_+^n)}{i!} \widehat{s}_i(x). \end{aligned} \tag{10.17}$$

Proof. The proof follows using Theorem 10.4 and Peano's theorem. □

From the theorem, classical error bounds can be obtained by applying well-known Hölder inequalities.

Now a new approximant polynomial can be considered.

Let z be a fixed point of the interval $[a,b]$. Then we get the polynomials

$$\begin{aligned} \widehat{P}_n[f,z](x) &:= f(z) + P_n[f](x) - P_n[f](z) = f(z) - \sum_{i=1}^{n} \frac{\mathcal{L}(Q^i f)}{i!}(s_i(x) - s_i(z)), \\ \widehat{\overline{P}}_n[f,z](x) &:= f(z) + \overline{P}_n[f](x) - \overline{P}_n[f](z) = f(z) - \sum_{i=1}^{n} \frac{\mathcal{L}(\widehat{Q}^i f)}{i!}(\widehat{s}_i(x) - \widehat{s}_i(z)). \end{aligned} \tag{10.18}$$

In the following, we will denote them by $\widehat{P}_n[f](x)$ and $\widehat{\overline{P}}_n[f](x)$ by omitting the dependence on z.

Theorem 10.6 ([64, 67]). *The polynomials $\widehat{P}_n[f](x)$ and $\widehat{\overline{P}}_n[f](x)$ are approximating polynomials of degree n for $f(x)$, that is, $\forall x \in [a,b]$,*

$$\begin{aligned} f(x) &= \widehat{P}_n[f](x) + \widehat{R}_n[f](x), \\ f(x) &= \widehat{\overline{P}}_n[f](x) + \widehat{\overline{R}}_n[f](x), \end{aligned} \tag{10.19}$$

with

$$\widehat{R}_n[x^i] = 0, \quad \widehat{\widehat{R}}_n[x^i] = 0, \quad i = 0, \dots, n,$$
$$\widehat{R}_n[x^{n+1}] \neq 0, \quad \widehat{\widehat{R}}_n[x^{n+1}] \neq 0. \tag{10.20}$$

Proof. The proof follows from the previous results. □

Theorem 10.7. *The polynomials $\widehat{P}_n[f](x)$ and $\widehat{\widehat{P}}_n[f](x)$ satisfy the interpolant conditions:*

$$\widehat{P}_n[f](z) = f(z), \quad \widehat{\widehat{P}}_n[f](z) = f(z),$$
$$\mathcal{L}(Q^i\widehat{P}_n[f]) = L(Q^i f), \quad \mathcal{L}(\widehat{Q}^i\widehat{\widehat{P}}_n[f]) = L(\widehat{Q}^i f), \quad i = 1, \dots, n. \tag{10.21}$$

We call $\widehat{P}_n[f](x)$ and $\widehat{\widehat{P}}_n[f](x)$ *the umbral interpolation polynomials of second kind* for $(\mathcal{L}, Q)$.

Remark 10.2. By using (10.10) and Remark 10.1, the umbral interpolant polynomial of second kind can be rewritten as

$$\widehat{P}_n[f](x) = f(z) - \sum_{i=0}^{n} \frac{\mathcal{L}(g^{(i)})}{i!}(s_i(x) - s_i(0)). \tag{10.22}$$

Remark 10.3. The polynomial $\widehat{P}_n[f](x)$ in (10.22) can be considered a generalization of the well-known Taylor and Newton interpolant polynomials. In fact, for

$$\mathcal{L}(f) = f(0), \quad Qf = Df,$$

we get the Taylor polynomial, whereas for

$$\mathcal{L}(f) = f(0), \quad Qf = \Delta f,$$

we have the Newton polynomial [76, p. 74].

If we suppose that $Q^i f \in X$, $\forall i \geq 0$, the series

$$s[f](x) = \sum_{i=0}^{\infty} \mathcal{L}(Q^i f)\frac{s_i(x)}{i!}$$

can be considered, and its convergence can be studied. This series can be called the generalized Taylor series or Sheffer–Taylor series.

10.2 Operators approximation theory

Like binomial [133, 134, 153, 154] and Appell sequences [112], also the more general polynomial sequences of Sheffer type have been applied in approximation theory, in finite and semiinfinite intervals [37, 143].

In fact, on the finite interval $[0,1]$, some results are in [72, 73, 152, 172]. In particular, the class of operators $\mathfrak{L}_n : C[0,1] \longrightarrow C[0,1]$, defined by

$$\mathfrak{L}_n[f] = \frac{1}{p_n(1)} \sum_{k=0}^{n} \binom{n}{k} s_k(x) p_{n-k}(1-x) f\left(\frac{k}{n}\right), \tag{10.23}$$

where $\{p_n\}_n$ and $\{s_n\}_n$ are b. p. s. and S. p. s. associated, is considered in [72], and the convergence properties are given.

In [152], there are the following results:

Let Q be a δ-operator with $\{p_n\}_n$ (the b. p. s. associated) and $\{s_n\}_n$ (the S. p. s. associated). There exists an invertible shift operator T such that $Ts_n(x) = p_n(x)$. Moreover, the following hypotheses are true:

1. The b.p.s. $\{p_n\}_n$ and the S.p.s $\{s_n\}_n$ satisfy: $p_n'(x) \geq 0$, $s_n(0) \geq 0$, $s_n(1) \neq 0$ $\forall n = 1, 2, \ldots$.
2. The numerical sequences

$$a_n = \frac{[(Q')^{-1} s_{n-1}](1)}{s_n(1)}, \qquad b_n = \frac{n-1}{n} \frac{[(Q')^{-2} s_{n-2}](1)}{s_n(1)},$$

where Q' is the Pincherle derivative [166, p. 49] of the operator Q, satisfy $\lim_{n\to\infty} a_n = \lim_{n\to\infty} b_n = 1$.

Then the author considers the operator sequence

$$P_n^{(Q,T)}[f](x) = \frac{1}{s_n(1)} \sum_{k=0}^{n} \binom{n}{k} p_k(x) s_{n-k}(1-x) f\left(\frac{k}{n}\right). \tag{10.24}$$

He proves the following:

Theorem 10.8. *With the previous notations and hypotheses, if $f \in C[0,1]$, then*

$$\lim_{n\to\infty} \|f - P_n^{(Q,T)}\| = 0, \tag{10.25}$$

where $\|f\| = \max_{0\leq t\leq 1} |f(t)|$.

We observe that if Q is the derivative operator and T is the identity, then the operator $P_n^{(Q,T)}[f]$ is the Bernstein operator, whereas if $(Q,T) = (Q,I)$, then it is the operator in (4.7).

In [172], a particular case of the operator (10.24) is considered.

For the semiinfinite interval, the operators introduced by Jakimovski [112] have been generalized to S. p. s. Ismail [110] considered the following operator sequence:

$$T_n[f](x) := \frac{\exp(-nx)H(1)}{A(1)} \sum_{k=0}^{\infty} s_k(nx) f\left(\frac{k}{n}\right), \tag{10.26}$$

where $\{s_k\}$ is the S. p. s. with the formal power series

$$A(t) = \sum_{k=0}^{\infty} a_k \frac{t^k}{k!}, \quad a_0 \neq 0,$$

$$H(t) = \sum_{k=1}^{\infty} b_k \frac{t^k}{k!}, \quad b_1 \neq 0.$$

To study the rate of convergence for this operator, in [189], the following hypotheses are considered:

1. $E = \{f : \forall x \in [0, +\infty[, |f(x)| \leq c\exp(kx), k \in \mathbb{R}, c > 0\}$;
2. for $x \in [0, +\infty[$, $k \in \mathbb{N}_0$, $s_k(x) \geq 0$;
3. $A(1) \neq 0$, $H'(1) = 1$;
4. the power series $A(t)$, $H(t)$ are analytic functions in the disc $|t| < R$, $R > 1$.

Under these hypotheses, the following statement is proved.

Theorem 10.9. *If $f \in \tilde{C}[0, \infty[\cap E$, for any $x \in [0, \infty[$, we have*

$$\left|T_n[f](x) - f(x)\right| \leq \left(1 + \sqrt{(H''(1) + 1)x + \frac{A'(1) + A''(1)}{nA(1)}}\right)\omega\left(f; \frac{1}{\sqrt{n}}\right), \tag{10.27}$$

given that $\tilde{C}[0, \infty[$ is the space of uniformly continuous functions, and $\omega(f, \delta)$ is the modulus of continuity of $f \in \tilde{C}[0, \infty[$.

Finally, also for operators (10.26), it is possible to apply the extrapolation algorithm to accelerate the convergence; related details has been considered in [55].

Theorem 10.10 ([45, 46, 55]). *Let $\{n_0 < n_1 < \cdots\}$ be an increasing sequence of positive integers, and let $h_i = n_i^{-1}$. We consider the sequence of polynomials of degree n_{i+q} defined by*

$$T_0^{(i)} := T_0^{(i)}(f)(x) = T_{n_i}[f](x) \tag{10.28}$$

$$T_q^{(i)} := T_q^{(i)}(f)(x) = \frac{h_{i+q}T_{q-1}^{(i)} - h_i T_{q-1}^{(i+1)}}{h_{i+q} - h_i}, \quad q = 1, \ldots, h-1.$$

Then

$$\lim_{h_i \to 0} T_q^{(i)} = f(x). \tag{10.29}$$

More precisely, the following representations of $T_q^{(i)}$ hold:

$$T_q^{(i)}(f)(x) = f(x) + h_i h_{i+1} \ldots h_q (-1)^q (c_{q+1}(f, x) + o(h_i)), \tag{10.30}$$

$$T_q^{(i)}(f)(x) = \sum_{j=0}^{q} l_j(0) T_{n_i}[f](x), \quad l_j(h) = \prod_{\substack{i \neq j \\ i=0}}^{q} \frac{h_i - h}{h_i - h_j}. \tag{10.31}$$

11 Examples

Abstract: With the methods of previous chapters, we consider some very important examples. In particular, we present generalized Bernoulli polynomials of the second kind, generalized Laguerre, generalized Boole, and Poisson–Charlier polynomials.

11.1 The generalized Bernoulli polynomials of the second kind

We consider the following:

1. The numerical sequences:
   –
   $$(\widehat{a}_i)_i \quad \text{with} \quad \widehat{a}_i := (-1)^i h^i \frac{i!}{i+1}, \quad i = 0, 1, \dots, \quad h > 0, \tag{11.1}$$
   $$(a_i)_i \quad \text{with} \quad \sum_{k=0}^{n} \binom{n}{k} a_i \widehat{a}_{n-i} = \delta_{n,0}, \tag{11.2}$$
   according to (2.10);
   –
   $$(\widehat{b}_i)_i \quad \text{with} \quad \widehat{b}_i := h^i := \widehat{b}_i(h), \quad i = 1, 2, \dots, \quad \widehat{b}_0 = 0, \tag{11.3}$$
   $$(b_i)_i \quad \text{with} \quad b_i = (-1)^{i+1}(i-1)!h^{i-1} := b_i(h), \quad i = 1, 2, \dots, \quad b_0 = 0, \tag{11.4}$$
   according to (2.7);
2. The conjugate Appell matrices:
   $$A := (a_{i,j}) \quad \text{with} \quad a_{i,j} = \binom{i}{j} a_{i-j}, \quad i, j = 0, \dots, i \geq j, \tag{11.5}$$
   $$\widehat{A} := (\widehat{a}_{i,j}) \quad \text{with} \quad \widehat{a}_{i,j} = \binom{i}{j} (-1)^{i-j} h^{i-j} \frac{(i-j)!}{i-j+1}, \quad i, j = 0, \dots, i \geq j, \tag{11.6}$$
3. The conjugate binomial matrices $P := (p_{i,j})$ and $\widehat{P} := (\widehat{p}_{i,j})$ as in Algorithms 2.1.1 and 2.1.2;
4. The conjugate Sheffer matrices $S := (s_{i,j})$ and $\widehat{S} := (\widehat{s}_{i,j})$, with
   $$s_{i,j} = \sum_{k=j}^{i} a_{i,k} p_{k,j} \quad \widehat{s}_{i,j} = \sum_{k=j}^{i} \widehat{p}_{i,k} \widehat{a}_{k,j}, \quad i = 0, 1, \dots, \quad j = 0, \dots, i, \tag{11.7}$$
   according to (2.26);

https://doi.org/10.1515/9783110652925-011

5. The power series:
   –

$$\frac{1}{g(t)} = \sum_{i=0}^{\infty} \widehat{a}_i \frac{t^i}{i!} = \sum_{i=0}^{\infty} (-1)^i h^i \frac{i!}{i+1} \frac{t^i}{i!} = \frac{ln(1+ht)}{ht}, \tag{11.8}$$

$$g(t) = \sum_{i=0}^{\infty} a_i \frac{t^i}{i!} = \frac{ht}{\ln(1+ht)}, \tag{11.9}$$

   –

$$\overline{f}(ht) = \sum_{i=0}^{\infty} \widehat{b}_i \frac{t^i}{i!} = \sum_{i=1}^{\infty} h^i \frac{t^i}{i!} = \exp(ht) - 1, \tag{11.10}$$

$$f(ht) = \sum_{i=1}^{\infty} b_i \frac{t^i}{i!} = \sum_{i=1}^{\infty} (-1)^{i+1} h^{i-1} \frac{t^i}{i} = \frac{1}{h} \ln(1+ht), \tag{11.11}$$

   –

$$Qy = \sum_{i=1}^{\infty} \widehat{b}_i \frac{y^{(i)}(x)}{i!} = \Delta_h y(x), \quad \widehat{Q}y = \sum_{i=1}^{\infty} (-1)^{i+1} h^{i-1} \frac{y^{(i)}(x)}{i}. \tag{11.12}$$

Then we get the conjugate S. p. s.:

$$\begin{array}{ll} s_0(x) = s_{0,0}, & \widehat{s}_0(x) = \widehat{s}_{0,0}, \\ s_1(x) = s_{1,0} + s_{1,1}x, & \widehat{s}_1(x) = \widehat{s}_{1,0} + \widehat{s}_{1,1}x, \\ \cdots\cdots\cdots & \cdots\cdots\cdots \\ s_n(x) = s_{n,0} + s_{n,1}x + \cdots + s_{n,n}x^n, & \widehat{s}_n(x) = \widehat{s}_{n,0} + \widehat{s}_{n,1}x + \cdots + \widehat{s}_{n,n}x^n, \\ \cdots\cdots\cdots & \cdots\cdots\cdots \end{array} \tag{11.13}$$

We note that (11.10) and (11.11) for $h = 1$ coincide with (5.8) and (5.9), which generate the conjugate b. p. s. called, respectively, exponential and lower factorial polynomial sequences. Consequently, we can consider the generalized lower factorial polynomials [113, p. 47]:

$$\begin{cases} (x)_0^h = h, \\ (x)_n^h = x(x-h)(x-2h)\cdots(x-(n-1)h), \quad n > 0, \quad h > 0, \end{cases} \tag{11.14}$$

and the generalized exponential polynomials $\phi_{n,h}(x)$ with

$$\begin{cases} \phi_{0,h}(x) = h, \\ \phi_{n+1,h}(x) = hx \sum_{k=0}^{n} \phi_{k,h}(x), \quad n \geq 0, \quad h > 0. \end{cases} \tag{11.15}$$

For simplicity, in the following, we write them as

$$(x)_n^h = (x)_n, \quad \phi_{n,h}(x) = \phi_n(x). \tag{11.16}$$

Then directly from the third determinantal form, we get

$$s_0(x) = 1,$$

$$s_n(x) = (-1)^{n+1} \begin{vmatrix} (x)_0 & (x)_1 & \dots & (x)_{n-1} & (x)_n \\ \widehat{a}_0 & \widehat{a}_1 & \dots & \widehat{a}_{n-1} & \widehat{a}_n \\ & \widehat{a}_0 & \dots & \binom{n-1}{1}\widehat{a}_{n-2} & \binom{n}{1}\widehat{a}_{n-1} \\ & & \ddots & \vdots & \vdots \\ & & & \widehat{a}_0 & \binom{n}{n-1}\widehat{a}_1 \end{vmatrix}, \tag{11.17}$$

and the recurrence relation

$$\begin{cases} s_0(x) = 1, \\ s_n(x) = (x)_n - n! \sum_{k=0}^{n-1} (-1)^{n-k} \frac{h^{n-k}}{(n-k+1)k!} s_k(x). \end{cases} \tag{11.18}$$

In analogy, for the conjugate

$$\widehat{s}_0(x) = 1,$$

$$\widehat{s}_n(x) = (-1)^{n+1} \begin{vmatrix} \phi_0(x) & \phi_1(x) & \dots & \phi_{n-1}(x) & \phi_n(x) \\ a_0 & a_1 & \dots & a_{n-1} & a_n \\ & a_0 & \dots & \binom{n-1}{1} a_{n-2} & \binom{n}{1} a_{n-1} \\ & & \ddots & \vdots & \vdots \\ & & & a_0 & \binom{n}{n-1} a_1 \end{vmatrix}, \tag{11.19}$$

and

$$\begin{cases} \widehat{s}_0(x) = 1, \\ \widehat{s}_n(x) = \phi_n(x) - \sum_{k=0}^{n-1} \binom{n}{k} a_{n-k} \widehat{s}_k(x). \end{cases} \tag{11.20}$$

For the first recurrence relation and determinantal form, we get

$$s_0(x) = s_{0,0}, \quad s_n(x) = \frac{1}{\widehat{s}_{n,n}} \left(x^n - \sum_{k=0}^{n-1} \widehat{s}_{n,k} s_k(x) \right), \tag{11.21}$$

$$\widehat{s}_0(x) = \widehat{s}_{0,0}, \quad \widehat{s}_n(x) = \frac{1}{s_{n,n}} \left(x^n - \sum_{k=0}^{n-1} s_{n,k} \widehat{s}_k(x) \right). \tag{11.22}$$

Moreover, it also holds that

$$s_n(x) = \frac{(-1)^n}{\prod_{i=0}^{n} \widehat{s}_{i,i}} \begin{vmatrix} 1 & x & \dots & x^{n-1} & x^n \\ \widehat{s}_{0,0} & \widehat{s}_{1,0} & \dots & \widehat{s}_{n-1,0} & \widehat{s}_{n,0} \\ & \ddots & & & \vdots \\ & & \ddots & & \vdots \\ & & & \widehat{s}_{n-1,n-1} & \widehat{s}_{n,n-1} \end{vmatrix}, \tag{11.23}$$

$$\widehat{s}_n(x) = \frac{(-1)^n}{\prod_{i=0}^n s_{i,i}} \begin{vmatrix} 1 & x & \dots & x^{n-1} & x^n \\ s_{0,0} & s_{1,0} & \dots & s_{n-1,0} & s_{n,0} \\ & \ddots & & & \vdots \\ & & \ddots & & \vdots \\ & & & s_{n-1,n-1} & s_{n,n-1} \end{vmatrix}. \tag{11.24}$$

Now let us consider the following sequences:

- (α_i) defined by

$$\sum_{k=0}^{n} \binom{n}{k} (-1)^{n-k+1} (n-k)! h^{n-k} \alpha_k = \delta_{n,0}; \tag{11.25}$$

- $(\widehat{\alpha}_i)$ defined by

$$\sum_{k=0}^{n} \binom{n}{k} h^{n-k+1} \widehat{\alpha}_k = \delta_{n,0}; \tag{11.26}$$

- (β_i) defined by

$$\beta_i = \sum_{k=0}^{n} \binom{n}{k} \alpha_k \left[\sum_{j=0}^{n-k} \binom{n-k}{j} (-1)^j h^{j-1} \frac{j!}{j+1} a_{n-k-j+1} \right]; \tag{11.27}$$

- $(\widehat{\beta}_i)$ defined by

$$\widehat{\beta}_i = \sum_{k=0}^{n} \binom{n}{k} \widehat{\alpha}_k \left[\sum_{j=0}^{n-k} \binom{n-k}{j} (-1)^{n-k-j+1} h^{n-k-j} \frac{(n-k-j+1)!}{n-k-j} a_k \right]. \tag{11.28}$$

Then we get the second recurrence relations and determinantal forms:

$$\begin{aligned}(n+1)\alpha_0 s_{n+1}(x) &= (x + \beta_0 - n\alpha_1) s_n(x) + \binom{n}{1} [\beta_1 - (n-1)\alpha_2] s_{n-1}(x) \\ &\quad + \binom{n}{2} [\beta_2 - (n-2)\alpha_3] s_{n-2}(x) + \dots + \beta_n s_0(x)\end{aligned} \tag{11.29}$$

and

$$\begin{aligned}(n+1)\widehat{\alpha}_0 \widehat{s}_{n+1}(x) &= (x + \widehat{\beta}_0 - n\widehat{\alpha}_1) \widehat{s}_n(x) + \binom{n}{1} [\widehat{\beta}_1 - (n-1)\widehat{\alpha}_2] \widehat{s}_{n-1}(x) \\ &\quad + \binom{n}{2} [\widehat{\beta}_2 - (n-2)\widehat{\alpha}_3] \widehat{s}_{n-2}(x) + \dots + \widehat{\beta}_n \widehat{s}_0(x).\end{aligned} \tag{11.30}$$

Moreover,

$$
\begin{aligned}
s_0(x) &= 1, \\
s_{n+1}(x) &= \frac{(-1)^{n+1}}{(\alpha_0)^{n+1}(n+1)!} \\
&\times \begin{vmatrix}
x+\beta_0 & -\alpha_0 & 0 & \cdots & 0 \\
\beta_1 & x+\beta_0-\alpha_1 & 1 & \cdots & 0 \\
\beta_2 & \binom{2}{1}[\beta_1-\alpha_2] & x+\beta_0-2\alpha_1 & \cdots & 0 \\
\vdots & & & \ddots & \vdots \\
\beta_{n-1} & \binom{n-1}{1}[\beta_{n-2}-\alpha_{n-1}] & \binom{n-1}{2}[\beta_{n-3}-2\alpha_{n-2}] & \cdots & -\alpha_0 \\
\beta_n & \binom{n}{1}[\beta_{n-1}-\alpha_n] & \binom{n}{2}[\beta_{n-2}-2\alpha_{n-1}] & \cdots & x+\beta_0-n\alpha_1
\end{vmatrix},
\end{aligned} \tag{11.31}
$$

$$
\begin{aligned}
\widehat{s}_0(x) &= 1, \\
\widehat{s}_{n+1}(x) &= \frac{(-1)^{n+1}}{(\widehat{\alpha}_0)^{n+1}(n+1)!} \\
&\times \begin{vmatrix}
x+\widehat{\beta}_0 & -\widehat{\alpha}_0 & 0 & \cdots & 0 \\
\widehat{\beta}_1 & x+\widehat{\beta}_0-\widehat{\alpha}_1 & 1 & \cdots & 0 \\
\widehat{\beta}_2 & \binom{2}{1}[\widehat{\beta}_1-\widehat{\alpha}_2] & x+\widehat{\beta}_0-2\widehat{\alpha}_1 & \cdots & 0 \\
\vdots & & & \ddots & \vdots \\
\widehat{\beta}_{n-1} & \binom{n-1}{1}[\widehat{\beta}_{n-2}-\widehat{\alpha}_{n-1}] & \binom{n-1}{2}[\widehat{\beta}_{n-3}-2\widehat{\alpha}_{n-2}] & \cdots & -\widehat{\alpha}_0 \\
\widehat{\beta}_n & \binom{n}{1}[\widehat{\beta}_{n-1}-\widehat{\alpha}_n] & \binom{n}{2}[\widehat{\beta}_{n-2}-2\widehat{\alpha}_{n-1}] & \cdots & x+\widehat{\beta}_0-n\widehat{\alpha}_1
\end{vmatrix}.
\end{aligned} \tag{11.32}
$$

From these determinantal forms, the polynomial $s_n(x)$ is the characteristic polynomial of a particular Hessenberg matrix, so the roots are the eigenvalues of this matrix.

It follows directly from the operator defined in (11.12) that

$$\Delta s_n(x) = n s_{n-1}(x), \tag{11.33}$$

whereas, after calculations ([63]), we get

$$D s_n(x) = n h(x)_{n-1}. \tag{11.34}$$

From (11.33) and (11.34), we can call the S. p. s. $\{s_n\}$ the *generalized Bernoulli polynomial of the second kind*, and denote it as $\{B^{\mathrm{II}}_{n,h}\}_n$. In fact, we get

$$B^{\mathrm{II}}_{n,h}(x) = n! h^{n+1} \Psi_n\left(\frac{x}{h}\right), \tag{11.35}$$

where $\Psi_n(x)$ is the Bernoulli polynomial of the second kind as defined in [113, p. 265]. The sequence $\{\widehat{s}_n\}$, instead, is not known in the literature; we call it the *generalized Sheffer exponential sequence* and denote by $\{S\phi_n\}_n$.

Now we list the first five polynomials of these sequences for $h = 1$:

$$\begin{aligned} B_0^{\mathrm{II}}(x) &= 1, & S\phi_0(x) &= 1,\\ B_1^{\mathrm{II}}(x) &= x + \frac{1}{2}, & S\phi_1(x) &= x - \frac{1}{2},\\ B_2^{\mathrm{II}}(x) &= x^2 - \frac{1}{6}, & S\phi_2(x) &= x^2 + \frac{1}{6},\\ B_3^{\mathrm{II}}(x) &= x^3 - \frac{3}{2}x^2 + \frac{1}{4}, & S\phi_3(x) &= x^3 + \frac{3}{2}x^2,\\ B_4^{\mathrm{II}}(x) &= x^4 - 4x^3 + 4x^2 - \frac{19}{30}, & S\phi_4(x) &= x^4 + 4x^3 + 2x^2 - \frac{1}{30}. \end{aligned}$$

Figures 11.1 and 11.2 show the graphics of these polynomials.

The numbers $B_n^{\mathrm{II}}(0)$ are called in the literature the *Cauchy numbers of the first type*, [42, 138], and it [113, p. 267] gives rise to

$$B_n^{\mathrm{II}}(0) = \int_0^1 (x)_n dx.$$

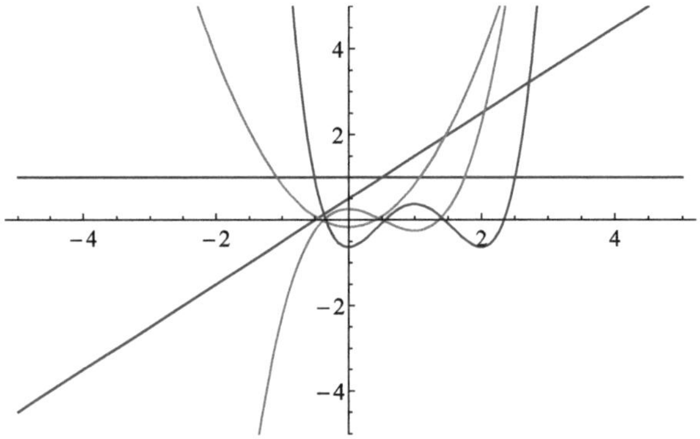

Figure 11.1: Generalized Bernoulli polynomials.

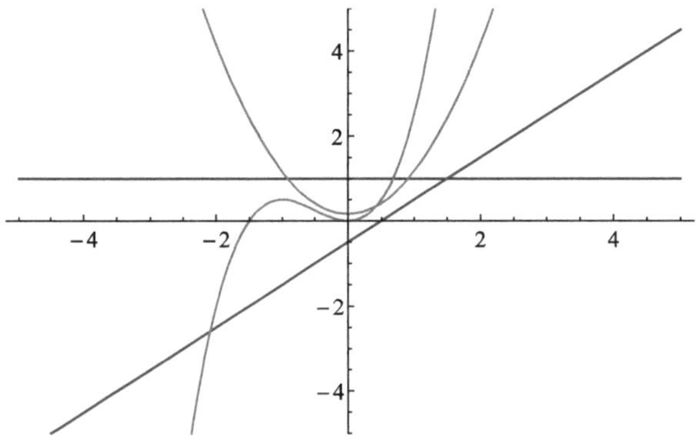

Figure 11.2: Sheffer exponential polynomials.

On the other hand, the numbers

$$b_n = \frac{1}{n!} B_n^{\mathrm{II}}(0) \tag{11.36}$$

are called in the literature the Bernoulli numbers of the second kind [166, p. 114].

The first six Bernoulli numbers of the second kind are

$$b_0 = 1, \quad b_1 = \frac{1}{2}, \quad b_2 = -\frac{1}{12}, \quad b_3 = \frac{1}{24}, \quad b_4 = -\frac{11}{720}, \quad b_5 = \frac{3}{160}.$$

The Bernoulli numbers of the second kind have been recently studied; see [145, 156–158, 160] and references therein.

We note explicitly that for $x = 0$, from (11.18), (11.21), (11.29), we get some recurrence relations for the Bernoulli numbers of the second kind:

$$b_n = -\sum_{k=0}^{n-1} \frac{(-1)^{n-k}}{n-k+1} b_k, \tag{11.37}$$

$$b_n = -\frac{1}{n!\widehat{s}_{n,n}} \sum_{k=0}^{n-1} k!\widehat{s}_{n,k} b_k, \tag{11.38}$$

$$(n+1)!\alpha_0 b_{n+1} = (\beta_0 - n\alpha_1)n!b_n + \binom{n}{1}[\beta_1 - (n-1)\alpha_2](n-1)!b_{n-1} + \cdots + b_n. \tag{11.39}$$

From (11.17), (11.23), and (11.31), for $x = 0$, we get the following determinantal forms:

$$b_n = \frac{(-1)^n}{n!} \begin{vmatrix} \frac{-1}{2} & \cdots & \frac{(-1)^{n-1}(n-1)!}{n} & \frac{(-1)^n n!}{n+1} \\ 1 & \cdots & \binom{n-1}{1}\frac{(-1)^{n-2}(n-2)!}{n-1} & \binom{n}{1}\frac{(-1)^{n-1}(n-1)!}{n} \\ & \ddots & & \vdots \\ & & 1 & -\binom{n}{n-1}\frac{1}{2} \end{vmatrix}, \tag{11.40}$$

$$b_n = \frac{(-1)^n}{n!\prod_{i=0}^{n}\widehat{s}_{i,i}} \begin{vmatrix} \widehat{s}_{1,0} & \cdots & \widehat{s}_{n-1,0} & \widehat{s}_{n,0} \\ \widehat{s}_{1,1} & \cdots & \widehat{s}_{n-1,1} & \widehat{s}_{n,1} \\ & \ddots & & \vdots \\ & & \widehat{s}_{n-1,n-1} & \widehat{s}_{n,n-1} \end{vmatrix}, \tag{11.41}$$

$$b_n = \frac{(-1)^{n+1}}{n!(\alpha_0)^{n+1}\prod_{i=0}^{n}(i+1)} \times \begin{vmatrix} \beta_0 & 1 & 0 & \cdots & 0 \\ \beta_1 & \beta_0 - \alpha_1 & 1 & \cdots & 0 \\ \beta_2 & \binom{2}{1}[\beta_1 - \alpha_2] & \beta_0 - 2\alpha_1 & \cdots & 0 \\ \vdots & & & \ddots & \vdots \\ \beta_{n-1} & \binom{n-1}{1}[\beta_{n-2} - \alpha_{n-1}] & \binom{n-1}{2}[\beta_{n-3} - 2\alpha_{n-2}] & \cdots & 1 \\ \beta_n & \binom{n}{1}[\beta_{n-1} - \alpha_n] & \binom{n}{2}[\beta_{n-2} - 2\alpha_{n-1}] & \cdots & \beta_0 - n\alpha_1 \end{vmatrix}. \tag{11.42}$$

The generating function of the generalized Bernoulli polynomials of the second kind is

$$G(x,t) = \frac{ht(1+ht)^{\frac{x}{h}}}{\ln(1+ht)}, \tag{11.43}$$

that is,

$$\frac{ht(1+ht)^{\frac{x}{h}}}{\ln(1+ht)} = \sum_{k=0}^{\infty} B_{k,h}^{\mathrm{II}}(x)\frac{t^n}{n!}. \tag{11.44}$$

For the generalized Sheffer exponential polynomials, we have

$$G_1(x,t) = \frac{\ln(1-h+h\exp(ht))}{h(\exp(ht)-1)}\exp(x(\exp(ht)-1)), \tag{11.45}$$

that is,

$$\frac{\ln(1-h+h\exp(ht))}{h(\exp(ht)-1)}\exp(x(\exp(ht)-1)) = \sum_{n=0}^{\infty} S\phi_{n,h}(x)\frac{t^n}{n!}. \tag{11.46}$$

For the Bernoulli polynomials of the second kind, we also have the differential equations (9.47) and (9.49).

The general properties of symmetry, extreme, roots of $\{B_{n,h}^{\mathrm{II}}\}_n$ are well studied, for example, in [113, pp. 268–270]. Therefore, we consider only the interpolation problem.

Let L be the linear functional defined by

$$L(f) = [D\Delta^{-1}f]_{x=0}, \tag{11.47}$$

where Δ^{-1} is the inverse of the difference operator Δ [113, p. 100], and after calculation, we get

$$L(x^i) = (-1)^i h^i \frac{i!}{i+1} = \hat{a}_i, \quad i = 0, 1, \dots. \tag{11.48}$$

Then, according to (10.11), the Bernoulli interpolation polynomial of the second kind of the first type is

$$B_n^{\mathrm{II}}[f](x) = h[\Delta^{-1}Df]_{x=0} + \sum_{i=1}^{n}\sum_{j=0}^{i-1}\frac{1}{h^i i!}\binom{i-1}{j}(-1)^{i-j-1}f'(jh)B_i^{\mathrm{II}}(x), \tag{11.49}$$

which, after calculation, can be expressed as

$$B_n^{\mathrm{II}}[f](x) = h[\Delta^{-1}Df]_{x=0} + \sum_{i=0}^{n} f'(ih)\mathcal{B}_i^{\mathrm{II}}(x), \tag{11.50}$$

where

$$\mathcal{B}_i^{\mathrm{II}}(x) = \sum_{j=i}^{n-1} \frac{1}{h^{j+1}} \binom{i}{j} \frac{(-1)^{j-1}}{(j+1)!} B_{j+1}^{\mathrm{II}}(x). \tag{11.51}$$

The polynomial $B_n^{\mathrm{II}}[f](x)$ satisfies the interpolant conditions

$$[\Delta^{-1} D B_n^{II}[f]]_{x=0} = [\Delta^{-1} Df]_{x=0}, \tag{11.52}$$

$$[\Delta^{i} D B_n^{II}[f]]_{x=0} = [\Delta^{i} Df]_{x=0}, \quad i = 0, \ldots, n-1. \tag{11.53}$$

Conditions (11.53) are equivalent to

$$B_n^{\mathrm{II}}[f]'(ih) = f'(ih), \quad i = 0, \ldots, n-1. \tag{11.54}$$

For the remainder, we have

$$R_n[f](x) = B_n^{\mathrm{II}}[f](x) - f(x). \tag{11.55}$$

Thus, if $f \in C^{n+1}[0, b]$ with $b = nh$, $\forall\, x \in\,]0, b[$, we get

$$R_n[f](x) = \frac{1}{n!} \int_0^b K_n(x, t) f^{(n+1)}(t) dt, \tag{11.56}$$

where

$$K_n(x, t) = (x - t)_+^n - nh[\Delta^{-1}(x - t)_+^{n-1}]_{x=0} - n \sum_{i=0}^{n} (ih - t)_+^{n-1} \mathcal{B}_i^{\mathrm{II}}(x). \tag{11.57}$$

Likewise, according to (10.18), the Bernoulli interpolant polynomials of the second kind are

$$\widehat{B}_n^{\mathrm{II}}[f](x) = f(0) + \sum_{i=0}^{n} f'(ih)(\mathcal{B}_i^{\mathrm{II}}(x) - \mathcal{B}_i^{\mathrm{II}}(0)), \tag{11.58}$$

with interpolant conditions

$$\widehat{B}_n^{\mathrm{II}}[f](0) = f(0), \quad \widehat{B}_n^{\mathrm{II}}[f]'(ih) = f'(ih), \quad i = 0, \ldots, n-1. \tag{11.59}$$

The interpolant polynomial $\widehat{B}_n^{\mathrm{II}}[f](x)$ suggests an application to the initial value problem

$$\begin{cases} y'(x) = f(x, y(x)), \\ y(0) = y_0, \end{cases} \quad x \in [0, b], \quad b > 0. \tag{11.60}$$

In fact, the polynomial

$$y_n(x) = y_0 + \sum_{i=0}^{n} f(ih, y_n(ih))(\mathcal{B}_i^{\mathrm{II}}(x) - \mathcal{B}_i^{\mathrm{II}}(0)) \tag{11.61}$$

is a collocation polynomial for problem (11.60) in the points $(ih, y_n(ih))$, $i = 0, \dots, n-1$, that is, it satisfies the condition

$$\begin{cases} y_n'(ih) = f(ih, y_n(ih)), & i = 0, \dots, n-1, \\ y_n(0) = y_0. \end{cases} \tag{11.62}$$

Implementation and numerical examples of the method (11.61) for IVP (11.60) are given in [49].

11.2 Generalized Laguerre polynomials

Laguerre polynomials are among the most important classical polynomial sequences in literature. These polynomials are used in every area of mathematics. In addition, they are used, also, as a solution of the Coulomb potential in quantum mechanics.

We consider the following:

1. The numerical sequences:
   -
     $$(a_n)_n \quad \text{with} \quad a_n := (1+\alpha)_n, \quad \alpha \in \mathbb{R}, \quad n \in \mathbb{N}, \tag{11.63}$$
     where
     $$(1+\alpha)_n = \begin{cases} 1 & \text{for} \quad \alpha = -1, \quad n \in \mathbb{N}, \\ 1 & \text{for} \quad \alpha \neq -1, \quad n = 0, \\ (1+\alpha)(2+\alpha)\cdots(n+\alpha) & \text{for} \quad \alpha \neq -1, \quad n > 0; \end{cases}$$
   -
     $$(b_n)_n \quad \text{with} \quad b_n := -n!, \quad n = 1, 2, \dots, \quad b_0 = 0. \tag{11.64}$$
2. The Appell matrix generated by the sequence (a_n), that is,
   $$A := (a_{i,j}) \quad \text{with} \quad a_{i,j} = \binom{i}{j} a_{i-j} = \frac{i!}{j!(i-j)!}(1+\alpha)(2+\alpha)\cdots(i-j+\alpha). \tag{11.65}$$
3. The binomial matrix generated by the sequence (b_n), that is,
   $$P := (p_{i,j}) \quad \text{with} \quad p_{i,j} = (-1)^j \frac{i!}{j!}\binom{i-1}{i-j} \tag{11.66}$$
   as in Algorithm 2.1.1 and according to (5.60);

4. The formal power series

$$g(t) = \sum_{n=0}^{\infty} a_n \frac{t^n}{n!} = \left(\frac{1}{1-t}\right)^{\alpha+1}, \tag{11.67}$$

$$f(t) = \sum_{n=0}^{\infty} b_n \frac{t^n}{n!} = \frac{t}{t-1}. \tag{11.68}$$

These are, respectively, an invertible and a δ-series;
5. The Sheffer matrix $S := AP$, that is, $S := (s_{i,j})$, where

$$s_{i,j} = \sum_{k=j}^{i} a_{i,k} p_{k,j} = (-1)^j \frac{i!}{j!} \sum_{k=j}^{i} \frac{1}{(i-k)!} (1+\alpha)_{i-k} \binom{k-1}{k-j} \tag{11.69}$$

or, equivalently, by the recursive formula

$$\begin{cases} s_{i,0} = (1+\alpha)_i, \\ s_{i,j+1} = \frac{1}{j+1} \sum_{k=j}^{i} \binom{i}{k} b_{i-k} s_{k,j}, \end{cases} \tag{11.70}$$

observing that

$$s_{i,i} = a_0 b_1^i = (-1)^i. \tag{11.71}$$

Then we get the S. p. s.

$$s_n(x) = \sum_{k=0}^{n} s_{n,k} x^k.$$

The first four polynomials of this sequence are:

$$s_0(x) = 1,$$
$$s_1(x) = 1 + \alpha - x,$$
$$s_2(x) = (1+\alpha)(2+\alpha) - 2(2+\alpha)x + x^2,$$
$$s_3(x) = (1+\alpha)(2+\alpha)(3+\alpha) - 3(2+\alpha)(3+\alpha)x + 3(3+\alpha)x^2 - x^3.$$

Directly from (9.22), we get the generating function of the S.p.s $\{s_n\}_n$:

$$\left(\frac{1}{1-t}\right)^{\alpha+1} \exp\left(\frac{xt}{t-1}\right) = \sum_{n=0}^{\infty} s_n(x) \frac{t^n}{n!}. \tag{11.72}$$

We note that the S. p. s. defined in (11.72) is the *generalized Laguerre polynomials* as defined in [166, p. 110], that is,

$$s_n(x) = L_n^{(\alpha)}(x), \quad \alpha \neq -1. \tag{11.73}$$

Many authors, including [20], [163, p. 201], take the generalized Laguerre polynomials to be $\frac{L_n^{(\alpha)}(x)}{n!}$.

We note that in [166, p. 109],

$$L_n^{(\alpha)}(x) = \sum_{k=0}^{n} \binom{n+\alpha}{n-k} \frac{n!}{k!}(-x)^k, \tag{11.74}$$

and that in [163, p. 201],

$$\frac{L_n^{(\alpha)}(x)}{n!} = \sum_{k=0}^{n} \frac{(-1)^k (1+\alpha)_n x^k}{k!(n-k)!(1+\alpha)_k}. \tag{11.75}$$

From the comparison with (11.69), we get the combinatorial identity

$$(-1)^k \frac{n!}{k!} \sum_{j=k}^{n} \frac{1}{(n-j)!}(1+\alpha)_{n-j} \binom{j-1}{j-k} = (-1)^k \frac{n!}{k!} \binom{n+\alpha}{n-k}. \tag{11.76}$$

The more well-known case of generalized Laguerre polynomials is for $\alpha = 0$; they are usually indicated as the Laguerre polynomials. The first four Laguerre polynomials are as follows:

$$\begin{aligned} L_0(x) &= 1, \\ L_1(x) &= 1 - x, \\ L_2(x) &= 2 - 4x + x^2, \\ L_3(x) &= 6 - 18x + 9x^2 - x^3, \end{aligned}$$

and their graphs are plotted in Figure 11.3.

We note that

$$\overline{f}(t) = f(t), \quad \frac{1}{g(\overline{f}(t))} = g(t), \tag{11.77}$$

and the conjugate S. p. s. $\{\widehat{s}_n\}_n$ coincides with $\{s_n\}_n$.

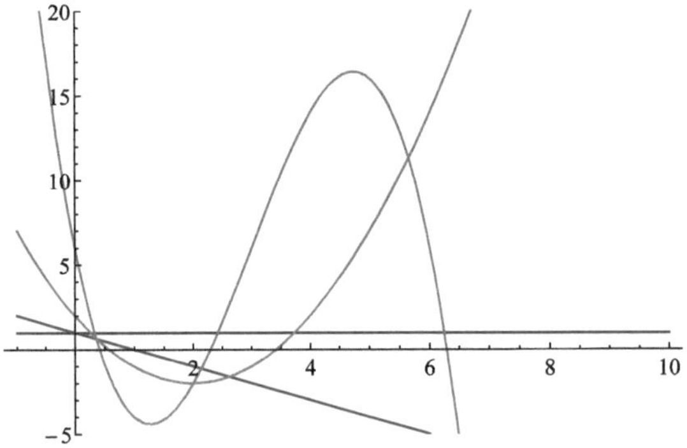

Figure 11.3: Laguerre polynomials.

By easy calculation, we get the following recurrence relations for $\{L_n^{(\alpha)}\}_n$:

$$L_n^{(\alpha)}(x) = \frac{1}{s_{n,n}}\left(x^n - \sum_{k=0}^{n-1} s_{n,k} L_k^{(\alpha)}(x)\right), \tag{11.78}$$

$$nL_n^{(\alpha)}(x) = (2n-1+\alpha-x)L_{n-1}^{(\alpha)}(x) - (n-1+\alpha)L_{n-2}^{(\alpha)}(x), \tag{11.79}$$

$$L_n^{(\alpha)}(x) = \frac{1}{\widehat{a}_{n,n}}\left(p_n(x) - \sum_{k=0}^{n-1}\binom{n}{k}\widehat{a}_{n-k}L_k^{(\alpha)}(x)\right). \tag{11.80}$$

We note also that from (5.62), we have $p_n(x) = n!L_n^{(-1)}(x)$, that is, $\{p_n\}_n$ is the b. p. s. for the δ-series $f(t)$, whereas $(\widehat{a}_n)$ is the conjugate sequence of (a_n), that is, the coefficients of the expansion of $\frac{1}{g(t)}$.

We observe that the recurrence (11.78) is equivalent to

$$x^n = \sum_{k=0}^{n} s_{n,k} L_k^{(\alpha)}(x). \tag{11.81}$$

Identity (11.81) allows us to expand any polynomial of degree less than n in generalized Laguerre polynomials. In fact, for $q_n(x) = \sum_{k=0}^{n} q_{n,k}x^k$, we get

$$q_n(x) = \sum_{k=0}^{n} c_{n,k} L_k^{(\alpha)}(x)$$

with

$$c_{n,k} = \sum_{j=k}^{n} q_{n,k}s_{j,k} = \sum_{j=k}^{n} q_{n,k}\frac{(-1)^k(1+\alpha)_j}{k!(j-k)!(1+\alpha)_k}.$$

The second recurrence relation (11.79) is a three-term relation. Therefore, from Remark 9.8, the generalized Laguerre polynomials for $\alpha \neq -1$ are orthogonal. In fact, it is proved [163, p. 205] that

$$\forall \alpha > -1 : \int_0^{+\infty} x^\alpha \exp(-x)L_n^{(\alpha)}(x)L_m^{(\alpha)}(x)dx = 0 \quad \text{for} \quad m \neq n. \tag{11.82}$$

We do not report all the properties of the generalized Laguerre polynomials that are derived from the orthogonality, but for them, we suggest [191].

The recurrence relation (11.80) is new and connects the general case $\alpha \neq -1$ with the particular case $\alpha = -1$.

Likewise, we get the determinantal forms:

$$L_n^{(\alpha)}(x) = (-1)^n \begin{vmatrix} 1 & x & \cdots & x^{n-1} & x^n \\ 1 & \widehat{s}_{1,0} & \cdots & \widehat{s}_{n-1,0} & \widehat{s}_{n,0} \\ & \ddots & & & \vdots \\ & & \ddots & & \vdots \\ & & & \widehat{s}_{n-1,n-1} & \widehat{s}_{n,n-1} \end{vmatrix}, \tag{11.83}$$

$$L_{n+1}^{(\alpha)}(x)=\frac{1}{(n+1)!}\begin{vmatrix} x-1-\alpha & 1 & 0 & \cdots & 0 \\ 1+\alpha & x-3-\alpha & 1 & \cdots & 0 \\ 0 & 3-\alpha & x-5-\alpha & \cdots & 0 \\ \vdots & & & \ddots & \vdots \\ 0 & \cdots & \cdots & & x-2n+1-\alpha \end{vmatrix}, \tag{11.84}$$

$$L_{n+1}^{(\alpha)}(x)=\frac{1}{(n+1)!}\begin{vmatrix} L_0^{(-1)}(x) & L_1^{(-1)}(x) & \cdots & (n-1)!L_{n-1}^{(-1)}(x) & n!L_n^{(-1)}(x) \\ \widehat{a}_0 & \widehat{a}_1 & \cdots & \widehat{a}_{n-1} & \widehat{a}_n \\ & \ddots & & & \vdots \\ & & \ddots & & \vdots \\ & & & \widehat{a}_0 & n\widehat{a}_1 \end{vmatrix}. \tag{11.85}$$

These three determinantal forms seem to be new. The first one can be useful for some determinantal identity and for the calculation of some particular values, considering that the Gauss method—without pivoting for Hessenberg matrix—is stable. The second form is very interesting. In fact, the generalized Laguerre polynomials are given as the characteristic polynomial of a tridiagonal matrix. Therefore, its roots can be calculated as the eigenvalues of the matrix. The third determinantal form can be useful for some determinantal identity.

Moreover, we have that the generalized Laguerre polynomials satisfy the following linear differential equation of the order n:

$$-ny(x)+\frac{1}{n}(x-(1+\alpha)+2(n-1))y'(x)-\left[\sum_{j=2}^{n}-\frac{j!}{n}(x-(1+\alpha)-2(n-1))\right.$$
$$\left.+\frac{1}{n}(1+\alpha+2(n-2))\right]\frac{y^{(j)}(x)}{j!}=0. \tag{11.86}$$

We now consider the linear functional $\mathcal{L}$ defined by

$$\mathcal{L}(f)=\int_0^{+\infty}\exp(-t)t^{\alpha}f(t)dt. \tag{11.87}$$

The results is

$$\mathcal{L}(x^n)=(1+\alpha)_n \quad \forall n\in\mathbb{N}. \tag{11.88}$$

It is an invertible linear functional. Then we consider the δ-functional M defined by

$$\begin{cases} M(x^k)=-\int_0^{+\infty}\exp(-t)t^k dt=-k!, & k=1,2,\ldots, \\ M(1)=0. \end{cases} \tag{11.89}$$

Then the generalized Laguerre polynomial $L_n^{(\alpha)}(x)$ satisfies the linear system

$$(\mathcal{L}M^k)(L_n^{(\alpha)}(x))=n!\delta_{n,k}, \quad n=1,2,\ldots, \quad k=0,\ldots,n. \tag{11.90}$$

The δ-operator associated with the generalized Laguerre polynomials is

$$Qy = -\sum_{i=1}^{\infty} i!\frac{y^{(i)}}{i!} = -\int_{0}^{+\infty} \exp(-t)y'(x+t)dt. \tag{11.91}$$

Other formulas involving the generalized Laguerre polynomials, obtained by umbral technique, are in [166, p. 109]. Therefore, we now consider the interpolation problem.

Let X be a linear space of real continuous functions defined in $[a, +\infty)$ with continuous derivatives of all necessary orders such that Qf exists for any $f \in X$. Moreover, let

$$XQ = \{f \in X \mid Q^i f \in X, i = 0, \dots, n, \forall n \in \mathbb{N}\}.$$

Let $\mathcal{L}$ be the linear functional on X defined in (11.87). Then from Theorems 10.1, 10.3, 10.4, 10.5 and Corollary 10.1, we get

$$\forall P_n \in \mathcal{P}_n \quad : \quad P_n(x) = \sum_{i=0}^{n}\sum_{k=0}^{i}(-1)^k \frac{i!}{k!}\binom{i-1}{k-1}\mathcal{L}(P_n^{(k)})\frac{L_i^{(\alpha)}(x)}{i!}, \tag{11.92}$$

$$\forall f \in XQ \quad : \quad f(x) = P_n[f](x) + R_n[f](x), \tag{11.93}$$

where

$$P_n[f](x) = \sum_{i=0}^{n}\sum_{k=0}^{i}(-1)^k \frac{i!}{k!}\binom{i-1}{k-1}\mathcal{L}(Q^k f)\frac{L_i^{(\alpha)}(x)}{i!}, \tag{11.94}$$

$$R_n[f](x) = \frac{1}{n!}\int_{0}^{+\infty} K_n(x,t)f^{(n+1)}(t)dt, \tag{11.95}$$

$$K_n(x,t) = R_n[(x-t)_+^n] = (x-t)_+^n - \sum_{i=0}^{n}\frac{\mathcal{L}(Q^i((x-t)_+^n))}{i!}L_i^{(\alpha)}(x). \tag{11.96}$$

Moreover, the following interpolant condition holds:

$$\mathcal{L}(Q^i f) = \mathcal{L}(Q^i P_n[f]), \quad i = 0, \dots, n. \tag{11.97}$$

The polynomial (11.92) can be considered as the expansion of a polynomial P_n in the generalized Laguerre polynomials, whereas the polynomial (11.94) can be considered a generalized Laguerre interpolant polynomial for any $f \in XQ$, with the remainder given by (11.95).

Finally, we can consider the generalized Laguerre interpolation of the second kind of $f \in XQ$:

$$f(x) = \overline{P}_n[f](x) + \overline{R}_n[f](x), \tag{11.98}$$

where

$$\overline{P}_n[f](x) = f(0) + \sum_{i=0}^{n}\sum_{k=0}^{i} \frac{(-1)^k}{k!}\binom{i-1}{k-1}\mathcal{L}(Q^k f)(L_i^{(\alpha)}(x) - (1+\alpha)_i), \tag{11.99}$$

$$\overline{R}_n[f](x) = \frac{1}{n!}\int_0^{+\infty} K(x,t) f^{(n+1)}(t)dt, \tag{11.100}$$

$$K(x,t) = (x-t)_+^n - \sum_{i=0}^{n} \frac{\mathcal{L}(Q^i((x-t)_+^n))}{i!}(L_i^{(\alpha)}(x) - (1+\alpha)_i). \tag{11.101}$$

The polynomial (11.99) can be considered as the generalized Laguerre–Taylor interpolation for f.

11.3 Generalized Boole polynomials

We consider the following:

1. The numerical sequences:
   –
   $$(a_i) \quad \text{with} \quad a_i := (-1)^i h^i \frac{i!}{2^i}, \quad i = 0, 1, \ldots, \quad h > 0, \tag{11.102}$$
   $$(\widehat{a}_i) \quad \text{with} \quad \widehat{a}_0 = 1, \quad \widehat{a}_1 = \frac{h}{2}, \quad \widehat{a}_i = 0, \quad i > 1, \tag{11.103}$$

   according to Algorithm 2.2.2;

   –
   $$(b_i) \quad \text{with} \quad b_i := (-1)^{i+1}(i-1)!h^{i-1}, \quad i = 1, 2, \ldots, \quad b_0 = 0, \tag{11.104}$$
   $$(\widehat{b}_i) \quad \text{with} \quad \widehat{b}_i = h^i, \quad i = 1, 2, \ldots, \quad \widehat{b}_0 = 0, \tag{11.105}$$

   according to Algorithm 2.1.2;
2. The conjugate Appell matrices

$$A := (a_{i,j}) \quad \text{with} \quad a_{i,j} = \binom{i}{j}(-1)^{i-j}h^{i-j}\frac{(i-j)!}{2^{i-j}}, \tag{11.106}$$

$$\widehat{A} := (\widehat{a}_{i,j}) \quad \text{with} \quad \widehat{a}_{i,j} = \binom{i}{j}\widehat{a}_{i-j} = \begin{cases} 1, & i = j, \\ \frac{h}{2}, & j = i-1, \\ 0, & j \neq i-1. \end{cases} \tag{11.107}$$

3. The conjugate binomial matrices $P := (p_{i,j})$ and $\widehat{P} := (\widehat{p}_{i,j})$ as in Algorithms 2.1.1 and 2.1.2, that is,

$$\begin{aligned} p_{n,0} &= \delta_{n,0}, \\ p_{n,1} &= (-1)^{n-1}(n-1)!, \\ p_{n,k} &= \frac{1}{k}\sum_{i=1}^{n-k+1}\binom{n}{i}(-1)^{i-1}(i-1)!p_{n-i,k-1}, \quad k = 2,\ldots, \end{aligned} \tag{11.108}$$

$$\begin{aligned} \widehat{p}_{n,0} &= \delta_{n,0}, \\ \widehat{p}_{n,1} &= h^i, \\ \widehat{p}_{n,k} &= \frac{1}{k}\sum_{i=1}^{n-k+1}\binom{n}{i}h^i\widehat{p}_{n-i,k-1}, \quad k = 2,\ldots. \end{aligned} \tag{11.109}$$

4. The conjugate Sheffer matrices $S := (s_{i,j})$ and $\widehat{S} := (\widehat{s}_{i,j})$, with

$$s_{i,j} = \sum_{k=j}^{i} a_{i,k}p_{k,j}, \quad \widehat{s}_{i,j} = \widehat{p}_{i,j} + \frac{(j+1)}{2}h. \tag{11.110}$$

5. The power series:
   –
$$g(t) = \sum_{i=0}^{\infty} a_i\frac{t^i}{i!} = \sum_{i=0}^{\infty}(-1)^i h^i\frac{i!}{2^i} = \frac{1}{1+\frac{h}{2}t}, \tag{11.111}$$
$$\frac{1}{g(t)} = \sum_{i=0}^{\infty}\widehat{a}_i\frac{t^i}{i!} = 1 + \frac{h}{2}t, \tag{11.112}$$
   –
$$f(ht) = \sum_{i=0}^{\infty} b_i\frac{t^i}{i!} = \sum_{i=1}^{\infty}(-1)^{i+1}h^{i-1}(i-1)!\frac{t^i}{i!} = \frac{1}{h}\ln(1+ht), \tag{11.113}$$
$$\bar{f}(ht) = \sum_{i=1}^{\infty}\widehat{b}_i\frac{t^i}{i!} = \sum_{i=1}^{\infty} h^i\frac{t^i}{i!} = \exp(ht) - 1. \tag{11.114}$$

Then we get the conjugate S. p. s.:

$$\begin{aligned} s_0(x) &= s_{0,0}, & \widehat{s}_0(x) &= \widehat{s}_{0,0}, \\ s_1(x) &= s_{1,0} + s_{1,1}x, & \widehat{s}_1(x) &= \widehat{s}_{1,0} + \widehat{s}_{1,1}x, \\ &\cdots\cdots & &\cdots\cdots \\ s_n(x) &= s_{n,0} + s_{n,1}x + \cdots + s_{n,n}x^n, & \widehat{s}_n(x) &= \widehat{s}_{n,0} + \widehat{s}_{n,1}x + \cdots + \widehat{s}_{n,n}x^n, \\ &\cdots\cdots & &\cdots\cdots \end{aligned} \tag{11.115}$$

We observe that, as in the generalized Bernoulli polynomials of the second kind, the conjugate b. p. s. associated with the numerical sequences (b_i), $(\widehat{b}_i)$ as given in (11.104),

(11.105), for $h = 1$, coincide with the conjugate b. p. s. called, respectively, falling- or factorial- and exponential polynomial sequences. Consequently, by considering the generalized falling polynomials $(x)_n^h \equiv (x)_n$, and the generalized exponential polynomials $\phi_{n,h}(x) = \phi_n(x)$, we get the conjugate S. p. s. (11.115) directly from the third determinantal form:

$$s_0(x) = 1,$$
$$s_n(x) = (-1)^{n+1} \begin{vmatrix} (x)_0 & (x)_1 & (x)_2 & \cdots & (x)_{n-1} & (x)_n \\ 1 & \frac{h}{2} & 0 & \cdots & 0 & 0 \\ 0 & 1 & \binom{2}{1}\frac{h}{2} & \cdots & 0 & 0 \\ & & & \ddots & & \vdots \\ 0 & & & \cdots & 1 & \binom{n}{n-1}\frac{h}{2} \end{vmatrix}, \tag{11.116}$$

and

$$\hat{s}_0(x) = 1,$$
$$\hat{s}_n(x) = (-1)^{n+1} \begin{vmatrix} (x)_0 & (x)_1 & (x)_2 & \cdots & (x)_{n-1} & (x)_n \\ a_0^* & a_1^* & a_2^* & \cdots & a_{n-1}^* & a_n^* \\ & & & \ddots & & \vdots \\ 0 & & & \cdots & a_0^* & \binom{n}{n-1}a_1^* \end{vmatrix}, \tag{11.117}$$

where $(a_{i,j}^*)$ are defined in (9.16).

We also have the recurrence relations

$$s_n(x) = (x)_n - n\frac{h}{2}s_{n-1}(x), \tag{11.118}$$

$$\hat{s}_n(x) = \phi_n(x) - \sum_{k=0}^{n-1} \binom{n}{k} a_{n-k}^* \hat{s}_k(x). \tag{11.119}$$

From (11.118), we can easily prove the relation

$$\begin{cases} \Delta s_n(x) = nhs_{n-1}(x), \\ s_n(0) = \frac{(-h)^n n!}{2^n}. \end{cases} \tag{11.120}$$

This relation in [113, p. 137], for $h = 1$ and up to a constant, characterizes the sequence of Boole polynomials. Therefore, we call them the *generalized Boole polynomials* and denote them by $\mathcal{B}_{n,h}(x)$ (with capital letter in honor of Boole). For $h = 1$, we set $\mathcal{B}_{n,1}(x) = \mathcal{B}_n(x)$. We can verify that

$$\mathcal{B}_{n,h}(x) = n!h^n \xi_n\left(\frac{x}{h}\right), \tag{11.121}$$

where $\xi_n(x)$ is the Boole polynomial considered in [113, p. 317].

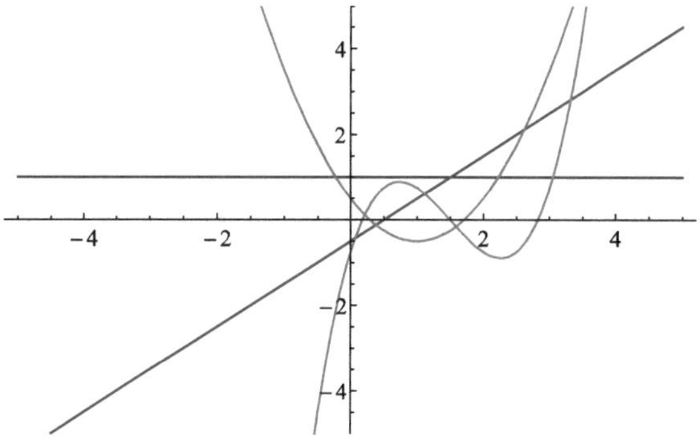

Figure 11.4: Generalized Boole polynomials.

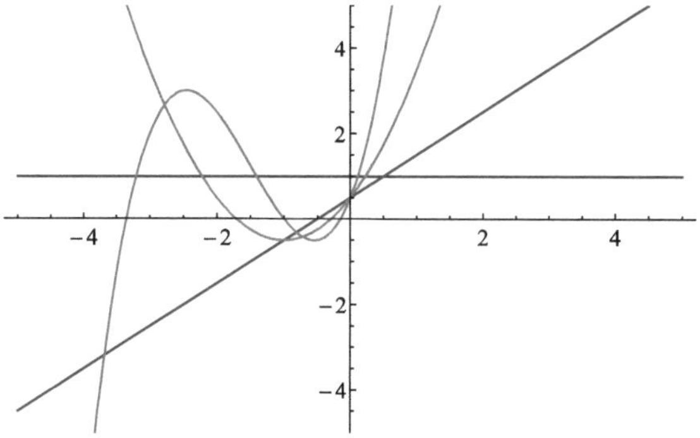

Figure 11.5: Generalized conjugate Boole polynomials.

The conjugate polynomial sequence $\{\widehat{s}_n\}_n$ does not seem to be present in the literature. We call it the *conjugate generalized Boole polynomials* and denote it by $\{\widehat{\mathcal{B}}_{n,h}\}$. The first polynomials for $h = 1$ are the following:

$$\begin{aligned}
\mathcal{B}_0(x) &= 1, & \widehat{\mathcal{B}}_0(x) &= 1,\\
\mathcal{B}_1(x) &= x - \frac{1}{2}, & \widehat{\mathcal{B}}_1(x) &= x + \frac{1}{2},\\
\mathcal{B}_2(x) &= x^2 - 2x + \frac{1}{2}, & \widehat{\mathcal{B}}_2(x) &= x^2 + 2x + \frac{1}{2},\\
\mathcal{B}_3(x) &= x^3 - \frac{9}{2}x^2 + 5x - \frac{3}{4}, & \widehat{\mathcal{B}}_3(x) &= x^3 + \frac{9}{2}x^2 + 4x + \frac{1}{2}.
\end{aligned}$$

In Figures 11.4 and 11.5, we show their graphs.

Directly from (11.111) and (11.115), we get the generating function of the generalized Boole polynomials

$$\frac{1}{1 + \frac{h}{2}t}\exp(x\ln(1 + ht)), \tag{11.122}$$

that is,

$$\frac{2}{2 + ht}(1 + ht)^{\frac{x}{h}} = \sum_{n=0}^{\infty} \mathcal{B}_{n,h}(x)\frac{t^n}{n!}, \tag{11.123}$$

and for its conjugate sequence, we get

$$\frac{1+\exp(ht)}{2}\exp(x(\exp(ht)-1)) = \sum_{n=0}^{\infty} \widehat{\mathcal{B}}_{n,h}(x)\frac{t^n}{n!}. \tag{11.124}$$

It is easy to determine other umbral properties of the generalized Boole polynomials. Therefore, in the following, we will study only the interpolant polynomial.

If M is the operator of mean [113, p. 6], that is,

$$Mf(x) = \frac{f(x)+f(x+h)}{2}, \tag{11.125}$$

we consider the linear functional

$$L(f) = [Mf]_{x=0}. \tag{11.126}$$

Then we have

$$\widehat{s}_i = \widehat{s}_{i0} = L(x^i), \quad i = 0,1,2,\dots. \tag{11.127}$$

Given the δ-operator for $\{\mathcal{B}_{n,h}\}_n$, the operator Δ_h, the Boole interpolant polynomial is

$$\mathcal{B}_{n,h}[f](x) = \sum_{i=0}^{n} \frac{[\Delta^i Mf]_{x=0}}{h^i i!}\mathcal{B}_{i,h}(x) = \sum_{i=0}^{n} \frac{f(ih)+f((i+1)h)}{2}\mathcal{E}_i(x), \tag{11.128}$$

where

$$\mathcal{E}_i(x) = \sum_{j=i}^{n} \binom{j}{i}\frac{(-1)^{j-i}}{h^j j!}\mathcal{B}_{n,h}(x), \quad i = 0,1,\dots. \tag{11.129}$$

The polynomial $\mathcal{B}_{n,h}[f](x)$ satisfies the interpolant conditions

$$\frac{f(ih)+f((i+1)h)}{2} = \frac{\mathcal{B}_{n,h}[f](ih)+\mathcal{B}_{n,h}[f]((i+1)h)}{2}, \quad i = 0,\dots,n. \tag{11.130}$$

Moreover, the second Boole interpolant polynomial is

$$\overline{\mathcal{B}}_{n,h}[f](x) = f(0) + \sum_{i=0}^{n} \frac{f(ih)+f((i+1)h)}{2}(\mathcal{E}_i(x)-\mathcal{E}_i(0)) \tag{11.131}$$

and satisfies

$$\overline{\mathcal{B}}_{n,h}[f](0) = f(0), \tag{11.132}$$

and, for $i = 1,\dots,n$,

$$[\Delta^i M\overline{\mathcal{B}}_{n,h}[f]]_{x=0} = [\Delta^i Mf]_{x=0}. \tag{11.133}$$

For the remainder, if $f \in C^{n+1}[0, b]$ with $b = nh$, $n > 0$, $\forall x \in [0, b]$, we get

$$R_n[f](x) = \frac{1}{n!} \int_0^b K_n(x,t) f^{(n+1)}(t)dt \tag{11.134}$$

with

$$K_n(x,t) = (x-t)_+^n - \frac{(-t)_+^n + (h-t)_+^n}{2} - \sum_{i=1}^{n} \sum_{j=0}^{i} \binom{i}{j} \frac{(-1)^{i-j}}{h^i j!} \frac{(jh-t)_+^n + ((j+1)h-t)_+^n}{2} \mathcal{B}_{i,h}(x). \tag{11.135}$$

Numerical examples for this interpolation are in [63].

The conjugate generalized Boole polynomial sequence $\{\widehat{\mathcal{B}}_{n,h}\}$ can be used in operators approximation theory. In fact, we get the following:

Proposition 11.1. *The polynomials $\{\widehat{\mathcal{B}}_{n,h}\}$ are positive for any $n \geq 0$ if $x \geq 0$.*

Proof. The proof follows from the second of (11.110), according to (11.105) and (11.109). □

Now, for $f \in \tilde{C}[0, \infty[\cap E$, we can consider the positive linear operators

$$\widehat{\mathcal{B}}_{n,h}[f](x) = \frac{\exp(-nx(\exp h - 1))}{1 + \frac{e}{2}} \sum_{k=0}^{\infty} \widehat{\mathcal{B}}_{k,h}(nx) f\left(\frac{k}{n}\right). \tag{11.136}$$

Then we get the following:

Theorem 11.1. *If $f \in \tilde{C}[0, \infty[\cap E$, then for any $x \in [0, \infty[$, we have*

$$|\widehat{\mathcal{B}}_{n,h}[f](x) - f(x)| \leq \left(1 + \sqrt{(h^2 e^2 + 1)x + \frac{h}{n(2+h)}}\right) \omega\left(f, \frac{1}{\sqrt{n}}\right). \tag{11.137}$$

Moreover, the extrapolation of Theorem 10.10 can be applied. A complete analysis of the operators (11.136) has been given in [55].

11.4 Poisson–Charlier polynomials

Poisson–Charlier polynomials have been discussed by Jordan [113, p. 473], Bateman [20, p. 276], Szëgo [191, p. 24], Roman [166, p. 119], and others.

With the method of previous examples, we can consider the following:

1. The numerical sequences:
   –

$$(a_n)_{n \in \mathbb{N}} \quad \text{with} \quad a_n = (-1)^n, \quad n = 0, 1, \ldots, \tag{11.138}$$

$$(\widehat{a}_n)_{n \in \mathbb{N}} \quad \text{with} \quad \widehat{a}_n = 1, \quad n = 0, 1, \ldots, \tag{11.139}$$

   according to Algorithm 2.2.2.

–

$$(b_n)_{n\in\mathbb{N}} \quad \text{with} \quad b_0 = 0, \quad b_n = a^{-n}(-1)^{n-1}(n-1)!, \quad n = 1, \dots \tag{11.140}$$

$$(\hat{b}_n)_{n\in\mathbb{N}} \quad \text{with} \quad \hat{b}_0 = 0, \quad \hat{b}_n = a, \quad n = 1, 2, \dots, \quad a \neq 0, a \in \mathbb{R}, \tag{11.141}$$

according to Algorithm 2.1.2.

We observe that the sequences b and $\hat{b}$ are related to sequences (5.6) and (5.7). Then the b. p. s. associated with these sequences are

$$p_n(x) = a^{-n}(x)_n = a^{-n}\sum_{k=0}^{n} s(n,k)x^k, \tag{11.142}$$

$$\hat{p}_n(x) = a\phi_n(x) = a\sum_{k=0}^{n} S(n,k)x^k, \tag{11.143}$$

where $s(n,k)$ and $S(n,k)$ are the Stirling numbers of the first and second kind.

2. The conjugate Appell matrices:

$$\begin{cases} A = (a_{i,j}) \quad \text{with} \quad a_{i,j} = \binom{i}{j}(-1)^{i-j}, \\ \hat{A} = (\hat{a}_{i,j}) \quad \text{with} \quad \hat{a}_{i,j} = \binom{i}{j}. \end{cases} \tag{11.144}$$

3. The binomial matrices generated by the sequences (b_i) and $(\hat{b}_i)$:

$$P = (p_{i,j}) \text{ with } \begin{cases} p_{l,0} = a^l \delta_{l,0}, & l = 0, 1, \dots, \\ p_{l,1} = b_l = a^{-l}(-1)^{l-1}(l-1)!, & l \geq 1, \\ p_{l,k} = \frac{1}{k}\sum_{i=1}^{l-k+1}(-1)^{i-1}\binom{l}{i}a^{-i}(i-1)!p_{l-i,k-1}, & l \geq k, \\ p_{l,k} = 0, & l < k, \end{cases}$$

$$\hat{P} = (\hat{p}_{i,j}) \text{ with } \begin{cases} \hat{p}_{l,0} = a^l \delta_{l,0}, & l = 0, 1, \dots, \\ \hat{p}_{l,1} = \hat{b}_l = a, & l \geq 1, \\ \hat{p}_{l,k} = \frac{1}{k}\sum_{i=1}^{l-k+1}\binom{l}{i}a\hat{p}_{l-i,k-1}, & k = 2, 3, \dots, \quad l \geq k, \\ \hat{p}_{l,k} = 0, & l < k, \end{cases} \tag{11.145}$$

as in Algorithm 2.1.1 and likewise (5.10)–(5.11).

From (11.143) and (11.142), it follows that the matrices P and $\hat{P}$ are the product of diagonal and Stirling matrices. That is,

$$P = D\mathrm{St}_1, \quad \hat{P} = D_1\mathrm{St}_2,$$

where $D = \mathrm{diag}\{\frac{1}{a^l}\}$, $D_1 = \mathrm{diag}\{a\}$, and St_1, St_2 are the Stirling matrices of the first and second kind.

For this, we can consider P and $\hat{P}$ as generalized Stirling's matrices.

4. The power series

$$\begin{cases} g(t) = \sum_{n=0}^{\infty}(-1)^n \frac{t^n}{n!} = \exp(-t), \\ \frac{1}{g(t)} = \sum_{n=0}^{\infty} \frac{t^n}{n!} = \exp(t), \end{cases} \tag{11.146}$$

$$\begin{cases} f(t) = \sum_{n=1}^{\infty} a^{-n}(-1)^{n-1} \frac{t^n}{n} = \log(1 + \frac{t}{a}), \\ \overline{f}(t) = a \sum_{n=1}^{\infty} \frac{t^n}{n!} = a(\exp(t) - 1). \end{cases} \tag{11.147}$$

5. The Sheffer matrices $S := AP$ and $\widehat{S} := \widehat{P}\widehat{A} \equiv \widehat{A}^*\widehat{P}$.

Then we get the conjugate S. p. s. $\{s_n\}_n$ and $\{\widehat{s}_n\}_n$ with

$$s_n(x) = \sum_{k=0}^{n} a_{n,k}p_k(x) = \sum_{j=0}^{n}\left(\sum_{k=0}^{n}\binom{n}{k}(-1)^{n-k}a^{-k}s(k,j)\right)x^j, \tag{11.148}$$

$$\widehat{s}_n(x) = \sum_{k=0}^{n} \widehat{a}^*_{n,k}\widehat{p}_k(x) = \sum_{k=0}^{n}\binom{n}{k}\widehat{a}^*_{n-k}a\sum_{j=0}^{k} S(k,j)x^k, \tag{11.149}$$

where $\widehat{A}^* := (\widehat{a}^*_{i,j})$ is the Appell matrix associated with the sequence $(\widehat{s}_i)$ defined in (2.18).

Remark 11.1. From (11.148) and (11.149), the matrices S and $\widehat{S}$ are, respectively,

$$S = (s_{i,j}) \quad : \quad s_{i,j} = \sum_{k=0}^{i}\binom{i}{k}(-1)^{i-k}a^{-k}s(k,j),$$

$$\widehat{S} = (\widehat{s}_{i,j}) \quad : \quad \widehat{s}_{i,j} = \binom{n}{k}\widehat{a}^*_{i-k}a\sum_{k=0}^{j} S(j,k).$$

The sequence $\{s_n\}_n$ is called the *Poisson–Charlier polynomials*, and it is indicated in [166, pp. 119–120] with $\{c_n(x,a)\}_n$, whereas the sequence $\{\widehat{s}_n\}_n$ is not known in the literature, and in the following it is indicated with $\{\widehat{c}_n\}$.

Other properties of these polynomials are known in the literature. Therefore, in the following, we consider only some recurrence, differential relations, and interpolant problems. For example, we have directly

$$c_n(x,a) = a^{-n}(x)_n - \sum_{k=0}^{n-1}\binom{n}{k}a^{n-k}c_k(x,a), \tag{11.150}$$

$$\widehat{c}_n(x,a) = \phi_n(ax) - \sum_{k=0}^{n-1}(-1)^{n-k}\binom{n}{k}\widehat{c}_k(x,a). \tag{11.151}$$

Likewise, we get the differential equations

$$\sum_{i=0}^{n}\frac{c_n^{(i)}(x,a)}{i!} = a^{-n}(x)_n, \quad \sum_{i=0}^{n}(-1)^n\frac{\widehat{c}_n^{(i)}(x,a)}{i!} = \phi_n(ax). \tag{11.152}$$

From (11.144) and (11.145), we get the generating function for the conjugate Poisson–Charlier sequences

$$\exp(-t)\left(1+\frac{t}{a}\right)^{x} = \sum_{n=0}^{\infty} c_n(x,a)\frac{t^n}{n!}, \tag{11.153}$$

$$\exp(t)\exp(x(a\exp(t)-1)) = \sum_{n=0}^{\infty} \widehat{c}_n(x,a)\frac{t^n}{n!}. \tag{11.154}$$

We now consider the δ-operators

$$Qy = \sum_{i=1}^{\infty} \widehat{b}_i \frac{y^{(i)}}{i!} = \sum_{i=1}^{\infty} a\frac{y^{(i)}}{i!} = \Delta_a y = a[y(x+1)-y(x)]. \tag{11.155}$$

From (11.155), we have the relation

$$\Delta_a c_n(x,a) = a(c_n(x+1,a)-c_n(x,a)) = nc_{n-1}(x,a) \tag{11.156}$$

and the functional equation

$$\frac{1}{n!}\Delta_a^n c_n(x,a) + \frac{\Delta_a^{n-1}}{(n-1)!}c_n(x,a) + \cdots + \Delta_a c_n(x,a) = a^{-n}(x)_n. \tag{11.157}$$

Finally, we can consider the Poisson–Charlier linear interpolation.

Let X be a linear space of real functions defined on $[a,b]$ with continuous derivatives of all necessary orders, and let L be a linear functional on X such that $L(1) \neq 0$. Then we get

$$\forall P_n(x) \in \mathcal{P}_n, \quad P_n(x) = \sum_{i=0}^{n} \frac{L(\Delta_a^i P_n(x))}{i!} c_i(x,a). \tag{11.158}$$

Given that

$$L(\Delta_a^i P_n) = L\left(a^i \sum_{k=0}^{i} \binom{i}{k} (-1)^k P_n(x+i-k)\right),$$

(11.158) becomes

$$P_n(x) = \sum_{i=0}^{n} a^i \sum_{k=0}^{i} \binom{i}{k} (-1)^k L(P_n(x+i-k)) \frac{c_i(x,a)}{i!}.$$

Likewise, for any $f \in X$ with $\Delta_a^i f \in X$, we get the interpolation polynomials

$$P_n([f])(x) = \sum_{i=0}^{n} \frac{L(\Delta_a^i f)}{i!} c_i(x,a), \tag{11.159}$$

which satisfy

$$L(\Delta_a^i f) = L(\Delta_a^i P_n[f]), \quad i = 0, \dots, n. \tag{11.160}$$

We also have the umbral polynomial interpolation of the second kind

$$\overline{P}_n[f](x) = f(z) - \sum_{i=0}^{n} \frac{L(\Delta_a^i f)}{i!}(c_i(x,a) - c_i(z,a)).$$

We note that the conjugate Poisson–Charlier polynomial sequence $\{\widehat{c}_n\}$ is positive for any $x \in [0, \infty[$ if $a > 0$. Then we can consider the positive linear operator

$$\widehat{c}_n[f,a](x) := \frac{\exp(-nx(a(e-1))}{e} \sum_{k=0}^{\infty} \widehat{c}_k(x,a) f\left(\frac{k}{n}\right). \tag{11.161}$$

We can apply to this operator the extrapolation as in Theorem 10.10. A more complete analysis of this operator is in [55].

Part V: **Lidstone polynomial sequences**

... It is proposed to show (by a simple line of direct algebraic demonstration which is also applicable to the original formula) that Aitkein's Theorem can be extended to the Everett type, i. c. the types which include two sets of term. One set involving $u(0)$ *and the resultant of generalized operator on* $u(0)$*, and the other set involving* $u(1)$ *and resultant of simulation operator on* $u(1)$ *...*
G. Lidstone (1930)

12 Lidstone-type polynomial sequences

Abstract: We consider a theory of Lidstone-type polynomials with the methods of the foregoing chapters. In particular, we present odd and even polynomial sequences, recurrence relations, and determinantal forms. A close relationship with Appell polynomials is presented. Finally, we sketch a relationship with linear functionals and linear interpolation.

12.1 Introduction

In 1929, Lidstone [128] introduced a polynomial expansion on two points by means of polynomial sequences connected to operator $D^2 = d^2/dx^2$. Later this expansion was characterized in terms of completely continuous functions in the papers of Boas [27], Portisky [155], Schoemberg [174, 175], Whittaker [205, 206], Widder [207, 208], and Golightly [97, 98]. In recent years, Agarwal et al. [6, 8, 9], Costabile et al. [47, 48, 70] connected the polynomial sequences of Lidstone to interpolatory problems [76] and to differential equations with boundary data. Lidstone polynomial sequences are connected to Bernoulli polynomials as written in [205], but they have never been framed in the modern umbral calculus.

Now we introduce an interesting relationship with A. p. s., which justifies their study in this context.

Moreover, Costabile et al. [70] introduced an algebraic approach to Lidstone polynomials, and in [47, 48], a first form of generalization of these sequences is proposed. In the following, therefore, there is an attempt to use a general algebraic approach to these topics.

12.2 Odd Lidstone-type polynomial sequences

Let O and $\widehat{O}$ be the matrices as in (2.27) and (2.29). That is, if (α_{2i}) with $\alpha_0 \neq 0$ and (β_{2i}) are numerical sequences such that

$$\sum_{i=0}^{k} \binom{2k+1}{2i+1} \frac{1}{2(k-j)+1} \alpha_{2i}\beta_{2(k-i)} = \begin{cases} 1, & k=0, \\ 0, & k \geq 1, \end{cases}$$

we get

$$O := (o_{i,j}) \quad \text{with} \quad \begin{cases} o_{i,0} = \alpha_{2i}, & i = 0, 1, \dots, \\ o_{i,j} = \binom{2i+1}{2j+1} \frac{\alpha_{2(i-j)}}{2(i-j)+1}, & i = 1, 2, \dots, \quad j = 1, \dots, i, \\ o_{i,j} = 0, & i < j, \end{cases} \tag{12.1}$$

https://doi.org/10.1515/9783110652925-012

and

$$\widehat{O} := (\widehat{o}_{i,j}) \quad \text{with} \quad \begin{cases} \widehat{o}_{i,0} = \beta_{2i}, & i = 0, 1, \ldots, \\ \widehat{o}_{i,j} = \binom{2i+1}{2j+1} \frac{\beta_{2(i-j)}}{2(i-j)+1}, & i = 1, 2, \ldots, \quad j = 1, \ldots, i, \\ \widehat{o}_{i,j} = 0, & i < j. \end{cases} \tag{12.2}$$

Then we can define the polynomials

$$\begin{aligned} &d_0(x) = o_{0,0}x, && \widehat{d}_0(x) = \widehat{o}_{0,0}x, \\ &d_1(x) = o_{1,0}x + o_{1,1}x^3, && \widehat{d}_1(x) = \widehat{o}_{1,0}x + \widehat{o}_{1,1}x^3, \\ &\cdots\cdots\cdots && \cdots\cdots\cdots \\ &d_k(x) = o_{k,0}x + \cdots + o_{k,k}x^{2k+1}, && \widehat{d}_k(x) = \widehat{o}_{k,0}x + \cdots + \widehat{o}_{k,k}x^{2k+1}, \\ &\cdots\cdots\cdots && \cdots\cdots\cdots \end{aligned} \tag{12.3}$$

These are called *conjugate odd Lidstone-type polynomial sequences* for $(\alpha_{2i})_i$ and $(\beta_{2i})_i$. From (12.1) and (12.2), we get

$$d_k(x) = \sum_{i=0}^{k} \binom{2k+1}{2i+1} \frac{\alpha_{2(k-i)}}{2(k-i)+1} x^{2i+1}, \tag{12.4}$$

$$\widehat{d}_k(x) = \sum_{i=0}^{k} \binom{2k+1}{2i+1} \frac{\beta_{2(k-i)}}{2(k-i)+1} x^{2i+1}. \tag{12.5}$$

It, therefore, follows that:
1. for any $k \in \mathbb{N}$, $d_k(x)$ and $\widehat{d}_k(x)$ are polynomials of degree $2k + 1$ with only odd degree monomials;
2. for any $k \in \mathbb{N}$, $d_k(x)$ and $\widehat{d}_k(x)$ have a zero in $x = 0$;
3. the coefficients α_{2i} and β_{2i}, $i = 0, \ldots, k$, in $d_k(x)$, $\widehat{d}_k(x)$ are independent by the degree.

In the following, we will denote with $\mathrm{OLP}(x)$ the set of odd Lidstone-type polynomials sequences, that is, of type (12.4) or (12.5).

Now we obtain some characterizations of the set $\mathrm{OLP}(x)$.

Proposition 12.1. *A polynomial set* $\{p_k\}$, $k = 0, 1, \ldots$, *is an element of* $\mathrm{OLP}(x)$ *if and only if the following statements hold:*
1. $p_0(x) \neq 0 \; \forall x \neq 0$,
2. $p_k(0) = 0 \; \forall k \in \mathbb{N}$,
3. $\text{degree}\, p_k(x) = 2k + 1 \; \forall k \in \mathbb{N}$,
4. $p_k''(x) = (2k + 1)(2k)p_{k-1}(x) \; \forall k = 1, 2, \ldots$.

Proof. The necessary condition easily follows from the definition. The sufficient condition follows by integrations of 4, according to 1 and 2. □

Remark 12.1. The set OLP(x) coincides, up to a factorial, with the set LG defined in [47].

Example 12.1. The sequence $\{x^{2k+1}\}_k$ is the OLP(x) related to the sequence $\alpha_0 = 1, \alpha_{2k} = 0$, $k > 0$. Other examples will be considered in Chapter 14.

Proposition 12.2. *For polynomial sequence $\{d_k\}_k \in \mathrm{OLP}(x)$, we get:*
1. $d_k(-x) = -d_k(x)$*-(symmetry with respect to the origin* $\forall k \in \mathbb{N}$*);*
2. $\int_0^1 d_k(x)dx = (2k+1)! \sum_{i=0}^{k} \frac{\alpha_{2(k-i)}}{(2i+2)!(2(k-i)+1)!}\ \forall\, k \in \mathbb{N}$;
3. $d_k^{(2j)}(x) = (2k+1)(2k)\cdots(2(k-j)+2)d_{k-j}(x), j = 1,\ldots,k \in \mathbb{N}$;
4. $d_k^{(2j+1)}(x) = (2k+1)(2k)\cdots(2(k-j)+2)d'_{k-j}(x), j = 1,\ldots,k \in \mathbb{N}$;
5. $d_k^{(2j)}(0) = 0, d_k^{(2j+1)}(0) = (2k+1)(2k)\cdots(2(k-j)+2)\alpha_{2(k-j)}, j = 1,\ldots,k, k = 1,2,\ldots,$ $d'_k(0) = \alpha_{2k}$;
6. $d_k(1) = 0 \Leftrightarrow \beta_0 = \beta_2 = \cdots = \beta_{2k}, k \geq 1$.

Proof. It follows using above definitions and propositions. See also [47, 57]. □

To determine the generating function of an OLP(x), we get the following:

Theorem 12.1 (Generating function). *A polynomial set $\{d_k\}_k$ is an element of* OLP(x) *if and only if there exists a sequence $(\alpha_{2k})_{k\in\mathbb{N}}$ with $\alpha_0 \neq 0$ such that setting*

$$l(t) = \sum_{k=0}^{\infty} \frac{\alpha_{2k}}{2k+1} \frac{t^{2k}}{(2k)!}, \tag{12.6}$$

we have

$$\frac{1}{t} \cdot l(t) \sinh tx = \sum_{k=0}^{\infty} \frac{d_k(x)}{2k+1} \frac{t^{2k}}{(2k)!}. \tag{12.7}$$

Proof. If $\{d_k\}_k \in \mathrm{OLP}(x)$, then there is a sequence $(\alpha_{2k})_k$ with $\alpha_0 \neq 0$ such that (12.4) holds. Then if

$$l(t) = \sum_{k=0}^{\infty} \alpha_{2k} \frac{t^{2k}}{(2k+1)!},$$

the result follows from the Cauchy series product.

Vice versa, if (12.6) and (12.7) hold, after the Cauchy series product, we get (12.4); hence the result. □

The function $\frac{l(t)}{t} \sinh tx$ is named the *generating function* of the sequence $\{d_k\}_k$.

Remark 12.2. If $\{d_k\}_k$ and $\{\widehat{d}_k\}_k$ are conjugate polynomial sequences of OLP(x) and if $\frac{l(t)}{t} \sinh tx$ is the generating function of $\{d_k\}_k$, then

$$t \cdot \frac{1}{l(t)} \sinh tx \tag{12.8}$$

is the generating function of $\{\widehat{d}_k\}_k$. We note that from (12.6), we have

$$\frac{1}{l(t)} = \sum_{k=0}^{\infty} \beta_{2k} \frac{t^{2k}}{(2k+1)!}, \tag{12.9}$$

with $(\beta_{2k})_k$ as

$$\sum_{i=0}^{k} \frac{\beta_{2i}\alpha_{2(k-i)}}{(2i+1)!(2(k-i)+1)!} = \begin{cases} 1, & k=0, \\ 0, & k \geq 1. \end{cases}$$

Also in the set $\mathrm{OLP}(x)$, we define the umbral composition.
Let it be that

$$d_k(x) = \sum_{i=0}^{k} o_{k,i} x^{2i+1} \quad \in \quad \mathrm{OLP}(x),$$

and

$$p_k(x) = \sum_{i=0}^{k} \overline{o}_{k,i} x^{2i+1} \quad \in \quad \mathrm{OLP}(x).$$

As a result, the umbral composition of $d_k(x)$ and $p_k(x)$ is

$$d_k \circ p_k = d_k(p_k(x)) := \sum_{i=0}^{k} o_{k,i} p_i(x), \tag{12.10}$$
$$p_k \circ d_k = p_k(d_k(x)) := \sum_{i=0}^{k} \overline{o}_{k,i} d_i(x).$$

Proposition 12.3. *If $\{d_k\}_k$, $\{\widehat{d}_k\}_k$ are conjugate polynomial sequences, respectively, for $(\alpha_{2i})_i$ and $(\beta_{2i})_i$, we get by umbral composition*

$$d_k(\widehat{d}_k(x)) = \widehat{d}_k(d_k(x)) = x^{2k+1}. \tag{12.11}$$

Proof. Using (12.10), we get

$$\widehat{d}_k(d_k(x)) = \sum_{i=0}^{k} \widehat{o}_{k,i} d_i(x) = \sum_{i=0}^{k} \widehat{o}_{k,i} \sum_{j=0}^{i} o_{i,j} x^{2j+1}$$
$$= \sum_{j=0}^{k} x^{2j+1} \sum_{i=j}^{k} \widehat{o}_{k,i} o_{i,j} = \sum_{j=0}^{k} x^{2j+1} \delta_{k,j} = x^{2k+1}.$$

□

Corollary 12.1. *The structure* $(\mathrm{OLP}(x), \circ)$ *is a group.*

Proof. It follows from (12.10), (12.11), and known results. We note that the identity is $\{x^{2i+1}\}_i$, and the inverse elements are the conjugate sequences. □

Theorem 12.2 (Relationship with Appell polynomial sequences). *The following statements hold:*
1. *If $\{a_k\}_k$ is an A. p. s., there exists a unique* OLP(x) $\{d_k\}_k$ *that is said to be associated with $\{a_k\}_k$;*
2. *If $\{d_k\}_k$ is in* OLP(x), *there exists an A. p. s. $\{a_k\}_k$ that is said to be associated with $\{d_k\}_k$;*
3. *If $\{a_k\}_k$ is an Appell sequence and we have $m \in \mathbb{R}$ such that*

$$a_{2k+1}\left(\frac{m}{2}\right) = 0, \quad \forall k \in \mathbb{N},$$

then the sequence

$$d_k(x) = 2^{2k+1} a_{2k+1}\left(\frac{x+m}{2}\right)$$

is the OLP(x) *associated.*

Proof.
1. Let $\{a_k\}_k$ be the A. p. s. generated by the numerical sequence (α_k) with $\alpha_0 \neq 0$. Then we consider the subsequence $(\alpha_{2k})_k$. This generates an OLP(x), $\{d_k\}$ that is said to be associated with $\{a_k\}_k$.
2. Let $\{d_k\}$ be the OLP(x) related to the numerical sequence (α_{2k}) with $\alpha_0 \neq 0$.We consider the numerical sequence $(\overline{\alpha}_i)_i$ such that

$$\overline{\alpha}_{2i} = \alpha_{2i}, \quad \overline{\alpha}_{2i+i} = \text{arbitrary} \quad \text{for} \quad i = 0, 1, \ldots.$$

   Then a sequence $(\overline{\alpha}_i)$ gives the A. p. s. $\{a_k\}_k$ that is said to be associated with $\{d_k\}$.
3. The proof follows using Proposition 12.1. It is also in [48]. □

12.3 Recurrence relation, determinantal form

For the conjugate odd Lidstone-type polynomial sequences, we can get the following recurrence relations:

Theorem 12.3 (Recurrence relation, [70]). *Two polynomial sequences $\{d_k\}_{k\in\mathbb{N}}$ and $\{\widehat{d}_k\}_{k\in\mathbb{N}}$ are conjugate odd Lidstone-type polynomial sequences if and only if the sequences $(\alpha_{2i})_{i\in\mathbb{N}}$, $(\beta_{2i})_{i\in\mathbb{N}}$ exist as in (2.28), and*

$$d_k(x) = \frac{1}{\beta_0}\left\{x^{2k+1} - \sum_{j=0}^{k-1}\binom{2k+1}{2j+1}\frac{\beta_{2(k-j)}}{2(k-j)+1}d_j(x)\right\}, \tag{12.12}$$

$$\widehat{d}_k(x) = \frac{1}{\alpha_0}\left\{x^{2k+1} - \sum_{j=0}^{k-1}\binom{2k+1}{2j+1}\frac{\alpha_{2(k-j)}}{2(k-j)+1}\widehat{d}_j(x)\right\}. \tag{12.13}$$

Proof. The proof follows from Proposition 12.3 and its corollary. □

Now the determinantal form follows:

Theorem 12.4 (Determinantal form, [70]). *Two polynomial sequences $\{d_k\}_{k\in\mathbb{N}}$, $\{\widehat{d}_k\}_{k\in\mathbb{N}}$ are conjugate odd Lidstone-type polynomial sequences if and only if two numerical sequences $(\alpha_{2i})_i$ and $(\beta_{2i})_i$ exist as in (2.28) and*

$$d_0(x) = \frac{1}{\beta_0}x,$$

$$d_k(x) = \frac{(-1)^k}{3!5!\cdots(2k-1)!\beta_0^{k+1}} \times \begin{vmatrix} x & x^3 & x^5 & \cdots & x^{2k-1} & x^{2k+1} \\ \beta_0 & \beta_2 & \beta_4 & \cdots & \beta_{2(k-1)} & \beta_{2k} \\ 0 & 3!\beta_0 & \frac{5!}{3!}\beta_2 & \cdots & \frac{(2k-1)!}{(2k-3)!}\beta_{2(k-2)} & \frac{(2k+1)!}{(2k-1)!}\beta_{2(k-1)} \\ \vdots & & \ddots & & \vdots & \vdots \\ \vdots & & & \frac{(2i+1)!}{(2i+1-2j)!}\beta_{2(i-j+1)} & \vdots & \vdots \\ \vdots & & & & \ddots & \vdots \\ 0 & \cdots & & \cdots & (2k-1)!\beta_0 & \frac{(2k+1)!}{3!}\beta_2 \end{vmatrix}, \tag{12.14}$$

and

$$\widehat{d}_0(x) = \frac{1}{\alpha_0}x,$$

$$\widehat{d}_k(x) = \frac{(-1)^k}{3!5!\cdots(2k-1)!\alpha_0^{k+1}} \times \begin{vmatrix} x & x^3 & x^5 & \cdots & x^{2k-1} & x^{2k+1} \\ \alpha_0 & \alpha_2 & \alpha_4 & \cdots & \alpha_{2(k-1)} & \alpha_{2k} \\ 0 & 3!\alpha_0 & \frac{5!}{3!}\alpha_2 & \cdots & \frac{(2k-1)!}{(2k-3)!}\alpha_{2(k-2)} & \frac{(2k+1)!}{(2k-1)!}\alpha_{2(k-1)} \\ \vdots & & \ddots & & \vdots & \vdots \\ \vdots & & & \frac{(2i+1)!}{(2i+1-2j)!}\alpha_{2(i-j+1)} & \vdots & \vdots \\ \vdots & & & & \ddots & \vdots \\ 0 & \cdots & & \cdots & (2k-1)!\alpha_0 & \frac{(2k+1)!}{3!}\alpha_2 \end{vmatrix}. \tag{12.15}$$

Proof. If $\{d_k\}_k$ and $\{\widehat{d}_k\}_k$ are conjugate OLP(x), we have $(\alpha_{2i})_i$ and $(\beta_{2i})_i$ such that (2.28) holds. Moreover, from (12.12), it follows that

$$\sum_{j=0}^{k} \frac{(2k+1)!}{(2j+1)!(2(k-j)+1)!}\beta_{2(k-j)}d_j(x) = x^{2k+1}, \quad k = 0,1,\ldots, \tag{12.16}$$

which can be considered as a triangular infinite linear system in the unknown $d_j(x)$. Solving the system of the first $k+1$ equations with Cramer's rule, we get the result.

Vice versa, from (12.14), expanding with respect to the last column [70], we have (12.12).

In analogy, we get (12.15). □

Remark 12.3. The determinantal forms (12.14) and (12.15) can also be deduced from the generating function. In fact, from $\frac{1}{t}\cdot l(t)\sinh tx = \sum_{k=0}^{\infty} d_k(x)\frac{t^{2k}}{(2k+1)!}$, we get

$$\sinh tx = t\cdot\frac{1}{l(t)}\cdot\sum_{k=0}^{\infty} d_k(x)\frac{t^{2k}}{(2k+1)!}.$$

Then after (12.9) and the Cauchy product, we get the infinite lower triangular system (12.16) in the unknown $d_k(x)$, $k = 0,1,\ldots$. Solving it with Cramer's rule, we get the result.

Remark 12.4. If $\{d_k\}_{k\in\mathbb{N}} \in \mathrm{OLP}(x)$, putting $y = 1-x$, we consider the polynomial

$$p_k(y) := d_k(y) = d_k(1-x), \tag{12.17}$$

and we get

$$p_k''(y) = d_k''(1-x) = (2k+1)(2k)d_{k-1}(1-x) = (2k+1)(2k)p_{k-1}(y). \tag{12.18}$$

So for $i = 0,1,\ldots,k$, we get

$$p_k^{(2i)}(0) = d_k^{(2i)}(1),\quad p_k^{(2i)}(1) = d_k^{(2i)}(0) = 0, \tag{12.19}$$

$$p_k^{(2i+1)}(1) = d_k^{(2i+1)}(0). \tag{12.20}$$

Finally, for (12.19), $\{p_k\}_k \in \mathrm{OLP}(x)$ if and only if $d_k(1) = 0$.

Now we will give the relationship with linear functional.

Let $\mathcal{P}_k$ be the span$\{x^{2i+1}\}_{i=0,\ldots,k}$, and let L be the linear functional on $\mathcal{P}_k$, defined by

$$L(x^{2i+1}) = \beta_{2i},\quad i = 0,1,\ldots,\quad \beta_0 \neq 0. \tag{12.21}$$

Then we can consider the OLP(x) $\{d_k\}$ defined in (12.14). We say that the sequence $\{d_k\}$ is related to linear functional L, and we denote it as $\{d_{L,k}\}_k$.

Theorem 12.5. *The following statements hold:*

1. *Setting*

$$L_i(x^{2j+1}) := L(D^{(2i)}(x^{2j+1})),\quad j = 0,1,\ldots,k,\quad i = 0,\ldots,k, \tag{12.22}$$

we have

$$L_i(d_{L,k}(x)) = (2k+1)!\delta_{i,k},\quad i = 0,\ldots,k, \tag{12.23}$$

that is, $d_{L,k}(x)$ is the solution of the linear interpolation problem

$$L_i(p_{2k+1}(x)) = (2k+1)!\delta_{i,k}, \quad i = 0, \dots, k.$$

2. *For any $Q_k(x) \in \mathcal{P}_k$, we have*

$$Q_k(x) = \sum_{i=0}^{k} \frac{L(Q_k^{(2i)})}{(2i+1)!} d_{L,i}(x). \tag{12.24}$$

Proof.
1. The proof follows using determinantal form (12.14) and the functional (12.22).
2. The result follows from 1. □

12.4 Even Lidstone-type polynomial sequences

Let E and $\hat{E}$ be the infinite lower triangular matrices defined in (2.34) and (2.36). That is, given the numerical sequence $(y_{2i})_i$, $y_0 \neq 0$, we have

$$E := (e_{i,j}) \quad \text{with} \quad \begin{cases} e_{i,0} = y_{2i}, & i = 0, 1, \dots, \\ e_{i,j} = \binom{2i}{2j} y_{2(i-j)}, & i = 1, 2, \dots, \quad j = 1, \dots, i, \\ e_{i,j} = 0, & i < j, \end{cases} \tag{12.25}$$

and, determining $(\delta_{2i})_i$ by (2.35), that is,

$$\sum_{j=0}^{i} \binom{2i}{2j} y_{2j} \delta_{2(i-j)} = \begin{cases} 1, & i = 0, \\ 0, & i > 0, \end{cases}$$

we get

$$\hat{E} := (\hat{e}_{i,j}) \quad \text{with} \quad \begin{cases} \hat{e}_{i,0} = \delta_{2i}, & i = 0, 1, \dots, \\ \hat{e}_{i,j} = \binom{2i}{2j} \delta_{2(i-j)}, & i = 1, 2, \dots, \quad j = 1, \dots, i, \\ \hat{e}_{i,j} = 0, & i < j. \end{cases} \tag{12.26}$$

Then we can consider the polynomial sets:

$$\begin{aligned} & t_0(x) = e_{0,0}, && \hat{t}_0(x) = \hat{e}_{0,0}, \\ & t_1(x) = e_{1,0} + e_{1,1}x^2, && \hat{t}_1(x) = \hat{e}_{1,0} + \hat{e}_{1,1}x^2, \\ & \cdots\cdots\cdots && \cdots\cdots\cdots \\ & t_k(x) = e_{k,0} + e_{k,1}x^2 + \cdots + e_{k,k}x^{2k}, && \hat{t}_k(x) = \hat{e}_{k,0} + \hat{e}_{k,1}x^2 + \cdots + \hat{e}_{k,k}x^{2k}, \\ & \cdots\cdots\cdots && \cdots\cdots\cdots \end{aligned} \tag{12.27}$$

These polynomial sets are called the *conjugate even Lidstone-type polynomials* for $(y_{2i})_{i\in\mathbb{N}}$, $(\delta_{2i})_{i\in\mathbb{N}}$.

Remark 12.5. It results directly from (12.25), (12.26), (12.27) that

$$t_k(x) = \sum_{j=0}^{k} e_{k,j} x^{2j} = \sum_{j=0}^{k} \binom{2k}{2j} \gamma_{2(k-j)} x^{2j}, \tag{12.28}$$

and

$$\hat{t}_k(x) = \sum_{j=0}^{k} \hat{e}_{k,j} x^{2j} = \sum_{j=0}^{k} \binom{2k}{2j} \delta_{2(k-j)} x^{2j}. \tag{12.29}$$

In the following, we will denote as ELP(x) the set of even Lidstone-type polynomials, that is, of type (12.28) or (12.29).

Remark 12.6. The set ELP(x) coincides, up to a constant, with the set LG* considered in [48].

Example 12.2. The polynomial sequence $\{x^{2k}\}_k$ is the ELP(x) related to the sequence $\gamma_0 = 1, \gamma_{2k} = 0, k = 1, \ldots$.

Now we get some characterizations of the set ELP(x).

Proposition 12.4. *A sequence* $\{p_k\}_k \in$ ELP(x) *if and only if*

$$\begin{cases} p_0(x) \neq 0, \\ \text{degree}\, p_k(x) = 2k, \\ p_k''(x) = 2k(2k-1)p_{k-1}(x), \\ p_k'(0) = 0. \end{cases} \tag{12.30}$$

Proof. The proof follows, after easy calculation, from the above definitions. □

Proposition 12.5. *For* $t_k(x) \in$ ELP(x), *we get:*

$$t_k(x) = t_k(-x) \quad (\textit{symmetry with respect to } y \textit{ axis } \forall k \in \mathbb{N}); \tag{12.31}$$

$$t_k^{(2i)}(x) = 2k(2k-1)\cdots(2(k-i)+1)t_{k-i}(x), \quad i = 1, \ldots, k; \tag{12.32}$$

$$t_k^{(2i+1)}(x) = 2k(2k-1)\cdots(2(k-i)+1)t_{k-i}'(x), \quad i = 1, \ldots, k-1; \tag{12.33}$$

$$t_k^{(2i+1)}(0) = 0, \quad k \in \mathbb{N}, \quad i = 0, \ldots, k; \tag{12.34}$$

$$t_k^{(2i)}(0) = 2k(2k-1)\cdots(2(k-i)+1)\gamma_{2(k-i)}, \quad i = 1, \ldots, k; \tag{12.35}$$

$$\int_0^1 t_k(x)dx = \sum_{i=0}^{k} \binom{2k}{2i} \frac{\gamma_{2(k-i)}}{(2i+1)}, \quad k \in \mathbb{N}; \tag{12.36}$$

$$(t_k(1) = 0, \quad k \geq 1) \Leftrightarrow (\delta_0 = \delta_2 = \cdots = \delta_{2k}). \tag{12.37}$$

Proof. It easily follows from the above propositions. □

Defining the umbral composition in ELP(x) as in (12.10), we get the following:

Proposition 12.6. *With umbral composition, we have*

$$t_k(\widehat{t}_k(x)) = \sum_{i=0}^{k} e_{k,i}\widehat{t}_i(x) = \widehat{t}_k(t_k(x)) = \sum_{i=0}^{k} \widehat{e}_{k,i}t_i(x) = x^{2k}. \tag{12.38}$$

Proof. The proof follows from characterization of $e_{k,i}$ and $\widehat{e}_{k,i}$. □

Proposition 12.7 (Generating function). *A polynomial sequence $\{t_k\}_{k\in\mathbb{N}}$ is an element of* ELP(x) *if and only if there exists a sequence $\{y_{2k}\}_k$ with $y_0 \neq 0$ such that defining the power series*

$$h(t) = \sum_{k=0}^{\infty} y_{2k}\frac{t^{2k}}{(2k)!}, \tag{12.39}$$

we have

$$h(t)\cosh tx = \sum_{k=0}^{\infty} t_k(x)\frac{t^{2k}}{(2k)!}. \tag{12.40}$$

Proof. If $\{t_k\}_k \in$ ELP(x), there exists a sequence $\{y_{2k}\}_k$ such that (12.28) holds. Then considering $h(t) = \sum_{k=0}^{\infty} y_{2k}\frac{t^{2k}}{(2k)!}$, (12.40) is verified in accordance with the Cauchy series product.

Vice versa, by differentiation of (12.40) with respect to the variable x, we get (12.30). □

The function $h(t)\cosh tx$ of the above proposition is called the *generating function* of the sequence $\{t_k\}_k$.

Remark 12.7. If $\{t_k\}_k$, $\{\widehat{t}_k\}_k$ are conjugate polynomial sequences of even Lidstone-type and $h(t)\cosh tx$ is the generating function of $\{t_k\}_k$, then $\frac{1}{h(t)}\cosh tx$ is the generating function of $\{\widehat{t}_k\}_k$.

Remark 12.8. If $\{t_k\}_{k\in\mathbb{N}} \in$ ELP(x), putting $y = 1 - x$, we consider the polynomials

$$\bar{t}_k(y) := t_k(y) = t_k(1-x), \tag{12.41}$$

and we get

$$\bar{t}_k''(y) = (2k)(2k-1)t_{k-1}(1-x) = (2k)(2k-1)\bar{t}_{k-1}(y). \tag{12.42}$$

So for $i = 0, 1, \dots, k$, we get

$$\bar{t}_k^{(2i)}(0) = t_k^{(2i)}(1), \quad \bar{t}_k^{(2i)}(1) = t_k^{(2i)}(0), \tag{12.43}$$

$$\bar{t}_k^{(2i+1)}(1) = t_k^{(2i+1)}(0), \quad \bar{t}_k^{(2i+1)}(0) = t_k^{(2i+1)}(1). \tag{12.44}$$

Finally, for the second of (12.44), $\{\bar{t}_k(y)\}_k \in$ ELP(x) if and only if $t_k'(1) = 0$.

Theorem 12.6 (Relationship with Appell polynomial sequence). *The following statements hold:*

1. *If $\{t_k\}_k$ is an* ELP(x), *there exists an A. p. s. $\{a_k\}_k$ that we regard as associated with $\{t_k\}_k$;*
2. *If $\{a_k\}_k$ is an A. p. s., there exists a unique $\{t_k\}_k \in$* ELP(x) *that we regard as associated with $\{a_k\}_k$;*
3. *If $\{a_k\}$ is an Appell sequence and there exists $m \in \mathbb{R}$ such that*

$$a_{2k+1}\left(\frac{m}{2}\right) = 0, \quad k = 0, 1, \dots,$$

then the sequence

$$t_k(x) = 2^{2k} a_{2k}\left(\frac{x+m}{2}\right)$$

is the ELP(x) *associated.*

Proof. The proof is analogous to Theorem 12.2. □

Remark 12.9. From an Appell polynomial sequence, we can obtain an odd and an even polynomial sequence of Lidstone type.

12.5 Recurrence relation, determinantal form

For the conjugate even Lidstone-type polynomial sequences the following relations hold.

Theorem 12.7 (Recurrence relation). *The polynomial sequences $\{t_k\}_k$ and $\{\hat{t}_k\}_k$ are conjugate even polynomial sequences of Lidstone type if and only if there are the sequences $\{\gamma_{2k}\}_k$, $\{\delta_{2k}\}_k$ as in* (2.34) *and* (2.35), *and*

$$t_k(x) = \frac{1}{\delta_0}\left(x^{2k} - \sum_{j=0}^{k-1}\binom{2k}{2j}\delta_{2(k-j)}t_j(x)\right), \tag{12.45}$$

$$\hat{t}_k(x) = \frac{1}{\gamma_0}\left(x^{2k} - \sum_{j=0}^{k-1}\binom{2k}{2j}\gamma_{2(k-j)}\hat{t}_j(x)\right). \tag{12.46}$$

Proof. It follows from (12.30). □

Theorem 12.8 (Determinantal form). *The polynomial sequences $\{t_k\}_k$ and $\{\hat{t}_k\}_k$ are conjugate even polynomial sequences of Lidstone type if and only if there are the sequences*

$(\gamma_{2k})_k$, $(\delta_{2k})_k$ *as in* (2.34) *and* (2.35), *and*

$$t_0(x) = \frac{1}{\delta_0},$$

$$t_k(x) = \frac{(-1)^k}{\delta_0^{k+1}} \begin{vmatrix} 1 & x^2 & x^4 & \cdots & x^{2k-2} & x^{2k} \\ \delta_0 & \delta_2 & \delta_4 & \cdots & \delta_{2k-2} & \delta_{2k} \\ 0 & \delta_0 & \binom{4}{2}\delta_2 & \cdots & \binom{2k-2}{2}\delta_{2(k-2)} & \binom{2k}{2}\delta_{2(k-1)} \\ \vdots & & \ddots & & \vdots & \vdots \\ \vdots & & & \ddots & \vdots & \vdots \\ 0 & \cdots & & \cdots & \delta_0 & \binom{2k}{2(k-1)}\delta_2 \end{vmatrix}, \tag{12.47}$$

and

$$\hat{t}_0(x) = \frac{1}{\gamma_0},$$

$$\hat{t}_k(x) = \frac{1}{\gamma_0^{k+1}} \begin{vmatrix} 1 & x^2 & x^4 & \cdots & x^{2k-2} & x^{2k} \\ \gamma_0 & \gamma_2 & \gamma_4 & \cdots & \gamma_{2k-2} & \gamma_{2k} \\ 0 & \gamma_0 & \binom{4}{2}\gamma_2 & \cdots & \binom{2k-2}{2}\gamma_{2(k-2)} & \binom{2k}{2}\gamma_{2(k-1)} \\ \vdots & & \ddots & & \vdots & \vdots \\ \vdots & & & \ddots & \vdots & \vdots \\ 0 & \cdots & & \cdots & \gamma_0 & \binom{2k}{2(k-1)}\gamma_2 \end{vmatrix}. \tag{12.48}$$

Proof. The proof is similar to that of Theorem 12.4. □

Remark 12.10. As in Remark 12.3, the determinantal forms (12.47) and (12.48) can be obtained by the generating functions.

Now let $\mathcal{P}_k^*$ be the span$\{x^{2i}\}_{i=0,\dots,k}$, and let L be the linear functional on $\mathcal{P}_k^*$ defined by

$$L(x^{2i}) = \delta_{2i}, \quad i = 0, 1, \dots, \quad \delta_0 \neq 0. \tag{12.49}$$

We can consider the ELP(x) as in (12.47), which we denote with $t_{k,L}(x)$. Then we get:

Proposition 12.8. *The following statements hold:*

1. *Setting*

$$L_i(x^{2j}) = L(D^{(2i)}x^{2j}), \quad i,j = 0, \dots, k, \quad k = 0, 1, \dots,$$

we have

$$L_i(t_{k,L}(x)) = (2k)!\delta_{i,k}, \quad i = 0, 1, \dots, k, \tag{12.50}$$

that is, $t_{k,L}(x)$ is the solution of the linear interpolation problem

$$L_i(p_{2k}(x)) = (2k)!\delta_{i,k}, \quad i = 0, 1, \dots, k.$$

2. *For any $Q_k(x) \in \mathcal{P}_k^*$, we have*

$$Q_k(x) = \sum_{i=0}^{k} \frac{L(D^{2i} Q_k(x))}{(2i)!} t_{i,L}(x). \tag{12.51}$$

12.6 Summary

In this chapter, we get a relationship between Lidstone-type polynomial sequences connected to the operator $D^2 := \frac{d^2}{dx^2}$ and the classical Appell polynomial sequences. That justifies and allows us an attempt of algebraic approach, based on matrix calculus, of Lidstone-type polynomial sequences, different from the classic operators and power series methods (see [9, p. 2], [8, 27, 31, 128, 132], and references therein). Matricial forms, recurrence relations, and determinantal form are determined. Finally, a representation theorem for polynomials is given.

13 Application to linear interpolation and operators approximation theory

Abstract: As in the previous chapters, here we consider some applications of Lidstone-type polynomial sequences to linear interpolation and operators approximation theory.

13.1 Odd Lidstone-type linear interpolation problem

The set OLP(x) can be a useful set of bases for a general linear interpolation problem [76]. In fact, let F be a linear functional on $\mathcal{C}^{2k+1}[0,1]$ such that $F(x) \neq 0$.

We consider the numerical sequence

$$\beta_{2i} = F(x^{2i+1}), \quad i = 0, 1, \ldots, \tag{13.1}$$

and we denote by $d_k^{\beta}(x)$ the odd Lidstone-type polynomial defined in (12.14), with (β_{2i}) as in (13.1). Then we get the following:

Theorem 13.1 ([47, p. 554]). *For a given set of values ω_i, $i = 0, \ldots, k$, with $\omega_i \in \mathbb{R}$, the polynomial*

$$P_{F,k}(x) = \sum_{i=0}^{k} \frac{\omega_i}{(2i+1)!} d_i^{\beta}(x) \tag{13.2}$$

is the only polynomial that satisfies the interpolatory conditions

$$F(P_{F,k}^{(2i)}) = \omega_i, \quad i = 0, \ldots, k. \tag{13.3}$$

Corollary 13.1. *For any $q_k(x) \in$ OLP(x), we have*

$$q_k(x) = \sum_{i=0}^{k} F(q_k^{(2i)}) \frac{d_i^{\beta}(x)}{(2i+1)!}. \tag{13.4}$$

Remark 13.1. Expansion (13.4) can be considered as the generalized Taylor polynomial of $q_k(x)$ in the set of odd Lidstone-type polynomials.

Now we consider a function $f \in \mathcal{C}^{2k+1}[0,1]$.

Theorem 13.2 ([48, 56]). *With the previous notations, for any $f \in \mathcal{C}^{2k+1}[0,1]$, the following relation holds:*

$$f(x) = P_{F,k}[f](x) + R_{F,k}[f](x), \tag{13.5}$$

https://doi.org/10.1515/9783110652925-013

where

$$P_{F,k}[f](x) = \sum_{i=0}^{k} F(f^{(2i)})\frac{d_i^{\beta}(x)}{(2i+1)!}, \tag{13.6}$$

$$R_{F,k}[f](x) = \sum_{j=0}^{k} f^{(2i)}(0)\left[x^{2j} - (2j)!\sum_{i=0}^{j}\frac{F[t^{2(j-i)}]}{(2(j-i))!}\frac{d_{t,i}(x)}{(2i+1)!}\right] + \int_0^1 K(x,t)f^{(2k+1)}(t)dt, \tag{13.7}$$

$$K(x,t) = R_{F,k}[(x-t)_+^{2k}]. \tag{13.8}$$

Moreover the following interpolant conditions hold

$$F[f^{(2i)}] = F([P_k^{(2i)}]), \quad i = 0,\ldots,k.$$

Proof. The result follows using Theorem 13.1 and Peano's lemma [76]. □

Theorem 13.3 ([48]). *With the previous notations, for any $f \in C^{(2k+1)}[0,1]$ and for any $z \in [0,1]$, the following relation holds:*

$$f(x) = \overline{P}_{F,k}[f](x) + \overline{R}_{F,k}[f](x), \tag{13.9}$$

where

$$\overline{P}_{F,k}[f](x) = f(z) + \sum_{i=0}^{k} F\left[\frac{f^{(2i)}}{(2i+1)!}\right](d_i^{\beta}(x) - d_i^{\beta}(z)), \tag{13.10}$$

$$\overline{R}_{F,k}[f](x) = R_{F,k}(x) - R_{F,k}(z). \tag{13.11}$$

Moreover, the following interpolant conditions are true:

$$\overline{P}_{F,k}[f](z) = f(z), \quad F(\overline{P}_{F,k}^{(2i)}[f]) = F(f^{(2i)}), \quad i = 1,\ldots,k. \tag{13.12}$$

Proof. It follows from previous results by means of standard techniques. □

Remark 13.2. It is possible to consider the polynomial

$$\hat{P}_k[f](x) = P_{F,k}[f](x) + \sum_{j=0}^{k} f^{(2i)}(0)\left[x^{2j} - (2j)!\sum_{i=0}^{j}\frac{F[t^{2(j-i)}]}{(2(j-i))!}\frac{d_{t,i}(x)}{(2i+1)!}\right] \tag{13.13}$$

and the related interpolation conditions. The details are in [56].

13.2 Odd Lidstone-type generalized Szasz operators

The application of the odd Lidstone-type polynomial sequences to operators approximation theory is not present in the literature. Therefore, it is a new field of research that will be debated in the future.

In fact, it is possible to consider the generalized Szasz-type operators [41]

$$L_n[f](x) = \frac{\sinh(-nx)}{l(1)} \sum_{k=0}^{\infty} d_k(nx) f\left(\frac{k}{n}\right) \tag{13.14}$$

under the hypothesis that $l(1) \neq 0$, $d_k(x) \geq 0\ \forall x \in [0, \infty[$, $\forall k \in \mathbb{N}$. It is useful to study the convergence for real functions.

13.3 Even Lidstone-type interpolation problem

The set ELP(x), like the set OLP(x), can be a useful set of bases for general linear interpolation problems.

Let L_k^* be the span$\{x^{2i}\}_{i=0,\dots,k}$, and let F be a linear functional on $C^{2k}[0,1]$ such that $F(1) \neq 0$. Then considering the sequence

$$\delta_{2i} = F(x^{2i}), \quad i \geq 0, \tag{13.15}$$

we get the even polynomial sequence $\{t_k^F\}_k$ defined by (12.28), with (y_i) that satisfies (2.35) with (δ_{2i}) as in (13.15). So we have the following:

Proposition 13.1 ([48, p. 187]). *For a given set of values $\omega_i \in \mathbb{R}$, $i = 0, \dots, k$, the polynomial*

$$p_k(x) = \sum_{i=0}^{k} \frac{\omega_i}{(2i)!} t_i^F(x) \tag{13.16}$$

is the unique polynomial of even degree less than or equal to $2k$ that satisfies the interpolatory conditions

$$F(p_k^{(2i)}(x)) = \omega_i, \quad i = 0, \dots, k. \tag{13.17}$$

Remark 13.3. If $q_k(x)$ is a polynomial of L_k^*, then

$$q_k(x) = \sum_{i=0}^{k} F(q_k^{(2i)}) \frac{t_i^F(x)}{(2i)!}.$$

That is, every polynomial in L_k^* has an analogue Taylor form in ELP.

This result can be extended to any $f \in C^{2k}[0,1]$ by adding a remainder term $R_k[f](x)$. In fact, with the above notation, if

$$\omega_i = F(f^{(2i)}), \quad i = 0, \dots, k,$$

then polynomial (13.16) becomes

$$p_k^F[f](x) = \sum_{i=0}^{k} F(f^{(2i)}) \frac{t_i^F(x)}{(2i)!} \tag{13.18}$$

and satisfies the interpolation conditions

$$F(p_k^{(2i)}[f]) = F(f^{(2i)}), \quad i = 0, \dots, k. \tag{13.19}$$

Thus, we can write

$$f(x) = p_k^F[f](x) + R_k[f](x), \tag{13.20}$$

and $p_k^F[f](x)$ can be called the ELP-interpolatory polynomial for f.

For the remainder $R_k[f](x)$, the following theorem holds:

Theorem 13.4 ([48, pp. 188]). *Let $f \in C^{2k}[0,1]$ be such that $f^{(2k+1)}$ exists. Then for any $z \in [0,1]$ for the remainder*

$$R_k[f](z) = f(z) - p_k[f](z),$$

we get

$$R_k[f](z) = \sum_{i=0}^{k-1} \frac{f^{(2i+1)}(0)}{(2i+1)!} \left[x^{2i+1} - (2i+1)! \sum_{j=0}^{i+1} F[x^{(2(i-j)+1)}] \frac{t_i^F(x)}{(2i)!} \right] + \int_0^1 K(x,t) f^{(2k+1)}(t)dt, \tag{13.21}$$

with $K(x,t) = R_k[(x-t)_+^{2k}]$.

Now it is possible to consider the polynomial

$$\widehat{P}_k^F[f](x) = P_k^F[f](x) + \sum_{i=0}^{k-1} \frac{f^{(2i+1)}(0)}{(2i+1)!} \left[x^{2i+1} - (2i+1)! \sum_{j=0}^{i+1} F[x^{(2(i-j)+1)}] \frac{t_i^F(x)}{(2i)!} \right] \tag{13.22}$$

and to study the interpolation conditions. The details are in [56].

Remark 13.4. The convergence of the series $\sum_{i=0}^{\infty} F(f^{(2i)}) t_i^F(x)$ can be studied with the technique used in [68, 97, 155, 174, 175, 205, 207, 208].

13.4 Even Lidstone-type generalized Szasz operators

Ciupa [41], in the context of Favard–Szasz operators involving Sheffer polynomial sequences, introduced the operators

$$P_n[f](x) = \frac{1}{\cosh(1)\cosh(nx)} \sum_{k=0}^{\infty} t_k(nx) f\left(\frac{2k}{n}\right) \quad \forall n \in \mathbb{N}, \tag{13.23}$$

where the sequence $\{t_k\}_k$ is generated by $\cosh z \cosh(zx)$. That is,

$$\cosh z \cosh(zx) = \sum_{k=0}^{\infty} t_k(x) z^{2k}. \tag{13.24}$$

It is easy to verify that

$$t_k(x) = \frac{(1+x)^{2k} + (1-x)^{2k}}{2(2k)!} \tag{13.25}$$

and that $\{(2k)!t_k\}_k$ is an even Lidstone-type polynomial sequence, but $\{t_k\}_k$ is not a Sheffer sequence. It is proved in [41] that under the hypotheses
1. $C[0,\infty)$ is the set of all real functions continuous on $[0,\infty)$,
2. $w_p(x) = \exp(-px)$, $x \geq 0, p > 0$,
3. $C_p = \{f \in C[0,\infty) : w_p f$ *is uniform continuous and bounded on* $[0,\infty)\}$,

the following theorem holds:

Theorem 13.5. *If $f \in C_p$, then for each $x \geq 0$, we have*

$$\lim_{n\to\infty} P_n[f](x) = f(x),$$

with uniform convergence in each interval $[0,a]$, $a > 0$.

This theorem suggests to consider the general operator

$$P_n^*[f](x) := \frac{1}{l(1)\cosh(nx)} \sum_{k=0}^{\infty} t_k(nx) f\left(\frac{2k}{n}\right), \tag{13.26}$$

where $l(t)$ is an analytic function with $l(1) \neq 0$, $\{t_k\}_k$ is the ELP(x) related to $l(t)$, such that $t_k(x) \geq 0$, $\forall x \geq 0$. It is possible to study the convergence of the sequence $(P_n^*[f](x))_n$ to the function $f \in C_p$.

13.5 Summary

Following the above method, the general linear interpolation problems, expressed in OLP(x) and ELP(x) bases, have been considered. The convergence of the series of interpolating polynomials is not considered. For operators approximation theory, it is noted that in the literature there is present only one example of Favard–Szasz-type operator involving ELP(x) sequences.

14 Examples

Abstract: We consider some relevant examples: the odd and even polynomial sequences associated to Bernoulli and Euler polynomial sequences. We obtain the classic Lidstone polynomial sequences (of the first and second type) known in the literature and two new sequences that seem to be unknown in the literature. Applications to linear interpolation problem and boundary value problem for high-order differential equation are also given.

14.1 First type Lidstone polynomials

It is known from (8.56) that the Bernoulli polynomials of odd degree have a zero at $x = \frac{1}{2}$, that is,

$$B_{2k+1}\left(\frac{1}{2}\right) = 0 \quad \forall k \in \mathbb{N}.$$

Then, from Theorem 12.2, we can consider the odd polynomials sequence $\{d_k\}_k$ defined by

$$d_k(x) = 2^{2k+1} B_{2k+1}\left(\frac{x+1}{2}\right), \tag{14.1}$$

where $B_k(x)$ is the Bernoulli polynomial of degree k.

By the well-known properties of Bernoulli polynomials, we have $d_k(0) = d_k(1) = 0$, $k \geq 1$, and therefore the polynomials $d_k(x)$ satisfy the conditions [57]

$$\begin{cases} d_k''(x) = (2k+1)(2k)d_{k-1}(x), \\ d_k(0) = d_k(1) = 0, \quad k \geq 1. \end{cases}$$

We also get that

$$d_k'(x) = (2k+1)2^{2k} B_{2k}\left(\frac{x+1}{2}\right), \tag{14.2}$$

and using Proposition 12.2/5, we have

$$\alpha_{2k} = d_k'(0) = (2k+1)2^{2k} B_{2k}\left(\frac{1}{2}\right). \tag{14.3}$$

Combining the previous relation with (8.57/2), that is,

$$B_{2k}\left(\frac{1}{2}\right) = (2^{1-2k} - 1)B_{2k}, \tag{14.4}$$

https://doi.org/10.1515/9783110652925-014

we get

$$\alpha_{2k} = (2k+1)2(1-2^{2k-1})B_{2k}, \quad k = 0, 1, \ldots. \tag{14.5}$$

These, for $k = 1, 2, \ldots$, are coefficients connected to the expansion of csch t. Then we have

$$\frac{1}{t}l(t) = \frac{1}{t}\sum_{k=0}^{\infty} \alpha_{2k}\frac{t^{2k}}{(2k+1)!} = \frac{1}{t} + \sum_{k=1}^{\infty} \frac{2(1-2^{2k-1})}{(2k)!}B_{2k}t^{2k-1} = \operatorname{csch} t, \quad |t| < \pi. \tag{14.6}$$

Consequently, we get

$$\frac{t}{l(t)} = \frac{1}{\operatorname{csch} t} = \sinh t = \sum_{k=0}^{\infty} \frac{t^{2k+1}}{(2k+1)!}, \tag{14.7}$$

from which, comparing with (12.9), we get

$$\beta_{2k} = 1 \quad \forall k \in \mathbb{N}. \tag{14.8}$$

Now we get the generating function of the polynomial sequence $\{d_k\}_k$ defined in (14.1):

$$\operatorname{csch} t \cdot \sinh tx = \sum_{k=0}^{\infty} \frac{d_k(x)}{2k+1}\frac{t^{2k}}{(2k)!}. \tag{14.9}$$

For the conjugate sequence $\{\widehat{d}_k\}_k$, we have the generating function

$$\sinh t \cdot \sinh tx = \sum_{k=0}^{\infty} \frac{\widehat{d}_k(x)}{2k+1}\frac{t^{2k}}{(2k)!}. \tag{14.10}$$

Hence, the conjugate sequence $\{\widehat{d}_k\}$ satisfies the conditions

$$\begin{cases} \widehat{d}_k''(x) = (2k+1)(2k)\widehat{d}_{k-1}(x), & k = 1, \ldots, \\ \widehat{d}_k(0) = 0, \quad \widehat{d}_k'(0) = 1. \end{cases} \tag{14.11}$$

The odd polynomial sequence $\{d_k\}_k$, up to a constant $(2k+1)!$, coincides with the so-called Lidstone polynomial sequence [128]. Effectively, Lidstone [128] called “Everett first type polynomials” the interpolant polynomials in the basis $\{\frac{1}{(2k+1)!}d_k\}_k$, whereas “Everett second type polynomials” are the interpolant polynomials with even sequences as a basis, which we will consider afterwards.

Lidstone (first type) polynomials are denoted in the literature as $\{\Lambda_k\}_{k\in\mathbb{N}}$, that is, $\Lambda_k(x) = \frac{d_k(x)}{(2k+1)!}$. The conjugate sequence $\{\widehat{d}_k\}_k$, as in (14.10), is not known in the literature, and in the following, we will denote the sequence $\{\frac{1}{(2k+1)!}\widehat{d}_k\}_k$ with $\{\widehat{\Lambda}_k\}_k$.

The following are the first six polynomials of these sequences:

$$\begin{aligned}
\Lambda_0(x) &= x,\\
\Lambda_1(x) &= -\frac{1}{6}x + \frac{x^3}{3!},\\
\Lambda_2(x) &= \frac{7}{360}x - \frac{1}{6}\frac{x^3}{3!} + \frac{x^5}{5!},\\
\Lambda_3(x) &= -\frac{31}{15120}x + \frac{7}{360}\frac{x^3}{3!} - \frac{1}{6}\frac{x^5}{5!} + \frac{x^7}{7!},\\
\Lambda_4(x) &= \frac{127}{604800}x - \frac{31}{15120}\frac{x^3}{3!} + \frac{7}{360}\frac{x^5}{5!} - \frac{1}{6}\frac{x^7}{7!} + \frac{x^9}{9!},\\
\Lambda_5(x) &= -\frac{511}{23950080}x + \frac{127}{604800}\frac{x^3}{3!} - \frac{31}{15120}\frac{x^5}{5!} + \frac{7}{360}\frac{x^7}{7!} - \frac{1}{6}\frac{x^9}{9!} + \frac{x^{11}}{(11)!},\\
\widehat{\Lambda}_0(x) &= x,\\
\widehat{\Lambda}_1(x) &= \frac{1}{3!}(x + x^3),\\
\widehat{\Lambda}_2(x) &= \frac{1}{5!}\left(x + \frac{10}{3}x^3 + x^5\right),\\
\widehat{\Lambda}_3(x) &= \frac{1}{7!}(x + 7x^3 + 7x^5 + x^7),\\
\widehat{\Lambda}_4(x) &= \frac{1}{9!}\left(x + 12x^3 + \frac{126}{5}x^5 + 12x^7 + x^9\right),\\
\widehat{\Lambda}_5(x) &= \frac{1}{11!}\left(x + \frac{55}{3}x^3 + 66x^5 + 66x^7 + \frac{55}{3}x^9 + x^{11}\right).
\end{aligned}$$

Thereafter, we compare the graphs of these polynomials in Figures 14.1 and 14.2.

We note that the sequences $\{\Lambda_k\}_k$ and $\{\widehat{\Lambda}_k\}_k$ are not conjugate sequences, that is,

$$\Lambda_k(\widehat{\Lambda}_k(x)) \neq x^{2k+1}.$$

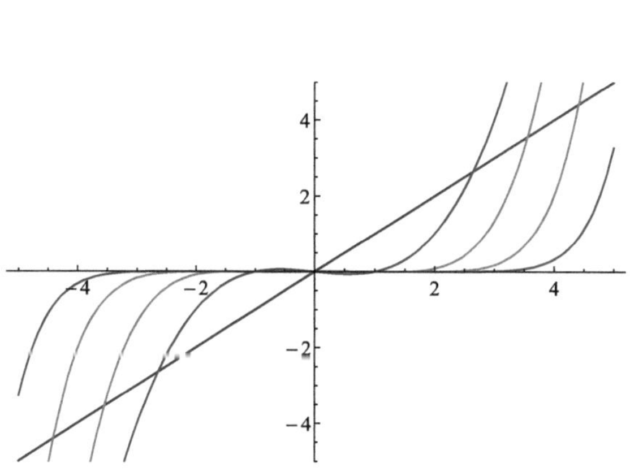

Figure 14.1: Odd Lidstone polynomials $\{\Lambda_k\}$.

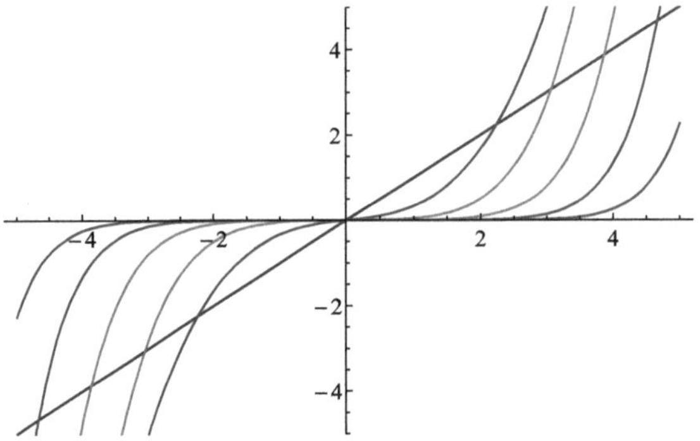

Figure 14.2: Odd Lidstone polynomials $\{\widehat{\Lambda}_k\}$.

The conjugate sequence of $\{\Lambda_k\}$ is the sequence $\{\overline{\Lambda}_k\}_k$ with

$$\overline{\Lambda}_k(x) = \sum_{i=0}^{k} \frac{(2k+1)!x^{2i+1}}{(2(k-i)+1)!}, \tag{14.12}$$

of which the first five polynomials are:

$$\begin{aligned}
\overline{\Lambda}_0 &= x, \\
\overline{\Lambda}_1 &= x + 3!x^3, \\
\overline{\Lambda}_2 &= x + \frac{5!}{3!}x^3 + 5!x^5, \\
\overline{\Lambda}_3 &= x + \frac{7!}{5!}x^3 + \frac{7!}{3!}x^5 + 7!x^7, \\
\overline{\Lambda}_4 &= x + \frac{9!}{7!}x^3 + \frac{9!}{5!}x^5 + \frac{9!}{3!}x^7 + 9!x^9.
\end{aligned}$$

We note that the sequence $\{\overline{\Lambda}_k\}$ does not satisfy any of the following differential equations:

$$\begin{aligned}
\overline{\Lambda}_k'' &= \overline{\Lambda}_{k-1}(x), \\
\overline{\Lambda}_k'' &= (2k+1)(2k)\overline{\Lambda}_{k-1}(x).
\end{aligned}$$

For some properties of the sequence $\{\widehat{\Lambda}_k\}_k$, we refer to [47], because the polynomial $\widehat{\Lambda}_k(x)$ satisfies the differential problem

$$\widehat{\Lambda}_k''(x) = \widehat{\Lambda}_{k-1}(x), \quad \widehat{\Lambda}_k(0) = 0.$$

In honor of Lidstone [128], in the following, we will consider some properties of the polynomial sequence $\{\Lambda_k\}_k$. The Lidstone sequence satisfies the recurrence relation [70]

$$\Lambda_k(x) = \frac{x^{2k+1}}{(2k+1)!} - \sum_{i=0}^{k-1} \frac{\Lambda_i(x)}{(2k-2i+1)!}, \quad k = 1, \dots, \tag{14.13}$$

and the determinantal form for $k = 1, 2, \dots$ is

$$\Lambda_k(x) = \frac{(-1)^k}{3!5!\cdots(2k+1)!}\begin{vmatrix} x & x^3 & x^5 & \cdots & x^{2k-1} & x^{2k+1} \\ 1 & 1 & 1 & \cdots & 1 & 1 \\ 0 & 3! & \frac{5!}{3!} & \cdots & \frac{(2k-1)!}{(2k-3)!} & \frac{(2k+1)!}{(2k-1)!} \\ \vdots & & \ddots & & \vdots & \vdots \\ \vdots & & & \ddots & \vdots & \vdots \\ 0 & \cdots & & \cdots & (2k-1)! & \frac{(2k+1)!}{3!} \end{vmatrix}. \tag{14.14}$$

To consider Lidstone interpolation [8, 9], we define the linear functional F on $C^{2k}[0,1]$ by

$$F(f) = f(1), \tag{14.15}$$

and then for $\{\Lambda_i\}_i$, we get:

$$\beta_{2i} = \frac{1}{(2i+1)!} = \frac{1}{(2i+1)!}F(x^{2i+1}). \tag{14.16}$$

For any $f \in C^{2k}[0,1]$, we consider the polynomial of degree $2k+1$

$$\Lambda_k[f](x) = \sum_{i=0}^{k} f^{(2i)}(1)\Lambda_i(x). \tag{14.17}$$

From Theorem 13.2, the interpolant conditions hold:

$$\Lambda_k^{(2i)}[f](1) = f^{(2i)}(1), \quad i = 0, \dots, k. \tag{14.18}$$

Moreover, from (13.5), we get:

$$f(x) = \Lambda_k[f](x) + R_k[f](x) \tag{14.19}$$

with [56]

$$R_k[f](x) = \sum_{i=0}^{k} f^{(2i)}(0)\Lambda_i(1-x) + \int_0^1 K(x,t)f^{(2k+2)}(t)dt, \tag{14.20}$$

where

$$K(x,t) = \frac{(x-t)_+^{2k+1}}{(2k+1)!} - \sum_{i=0}^{k} \frac{(1-t)^{2(k-i)+1}}{(2(k-i)+1)!}\Lambda_i(x).$$

The interpolation polynomial (14.17), written by means of the sequence $\{d_k\}_k$, assumes the form

$$d_k[f](x) = \sum_{i=0}^{k} \frac{f^{(2i)}(1)}{(2i)!}\frac{d_i(x)}{2i+1},$$

and it can be considered as a modified Taylor polynomial with only even derivatives.

From Remark 12.4 or from (14.20), we can consider the polynomial

$$L_k[f](x) = \sum_{i=0}^{k-1} [f^{(2i)}(1)\Lambda_i(x) + f^{(2i)}(0)\Lambda_i(1-x)]; \tag{14.21}$$

it satisfies the interpolation conditions

$$f^{(2i)}(1) = L^{(2i)}[f](1), \quad f^{(2i)}(0) = L^{(2i)}[f](0), \quad i = 0, \dots, k-1. \tag{14.22}$$

This polynomial is known as the *Lidstone interpolation polynomial* [9] but is called by Lidstone [128] the *Everett first type formula*.

For the remainder, it is known that for any $f \in C^{2k}[0,1]$, we have [9, 210, 211]

$$f(x) = L_k[f](x) + R_k[f](x), \tag{14.23}$$

with the error $R_k[f](x)$ expressed in the Cauchy form

$$R_k[f](x) = E_{2k}(x)f^{(2k)}(\xi_x), \quad \xi_x \in (0,1),$$

where $E_k(x)$ is the Euler polynomial of degree k.

Lidstone first type interpolation polynomial is used for the study of the so-called Lidstone boundary value problem [6, 8]:

$$\begin{aligned}
&(-1)^k y^{(2k)}(x) = f(x, \overline{y}(x)), \quad \overline{y}(x) = (y(x), y'(x), \dots, y^{(2k-1)}(x))^T, \\
&y^{(2i)}(0) = \alpha_i, \quad y^{(2i)}(1) = \beta_i, \quad i = 0, \dots, k-1.
\end{aligned} \tag{14.24}$$

For a collocation method for this problem, which uses Lidstone interpolation polynomials, we refer to [50].

Finally, the Lidstone series

$$\sum_{i=0}^{\infty} [f^{(2i)}(0)\Lambda_i(1-x) + f^{(2i)}(1)\Lambda_i(x)]$$

has been considered by many authors (see [27, 68, 97, 98, 173, 205, 208] and references therein). In particular, in [68] there is a new proof of uniform convergence for functions of exponential type. Also the spline interpolation with Lidstone boundary condition (14.22) is considered in [209–211]. A collocation spline Lidstone-based for BVP (14.24) with $k = 1$ is in [58], whereas the general case for $k > 1$ will be presented in a separate paper.

Other examples of odd Lidstone first type polynomial sequences are in [47]. In particular, we note formula (24) in [47, p. 548], which gives an expansion in Bernoulli polynomials of odd degree, whereas Example 3.3 in [47, p. 547] will be reconsidered with a new prospective.

14.2 Odd Lidstone–Euler polynomials

It is known [113, p. 292] that the Euler polynomials of odd degree have a zero in $x = \frac{1}{2}$, that is,

$$E_{2n+1}\left(\frac{1}{2}\right) = 0, \quad n = 0, 1, \dots .$$

Then, from Theorem 12.2, we can consider the odd polynomials sequence $\{d_k\}_k$ defined by

$$d_k(x) = 2^{2k+1} E_{2k+1}\left(\frac{x+1}{2}\right), \tag{14.25}$$

where $E_k(x)$ is the Euler polynomial of degree k.

We get

$$d_k'(x) = (2k+1)2^{2k} E_{2k}\left(\frac{x+1}{2}\right), \tag{14.26}$$

from which, recalling that $E_{2k}(1) = 0$ [113, p. 212], we get

$$d_k'(1) = 0.$$

Then according to (14.1), $\{d_k\}_k$ satisfies

$$\begin{cases} d_k''(x) = (2k+1)(2k)d_{k-1}(x), \\ d_k(0) = 0, \quad d_k'(1) = 0. \end{cases} \tag{14.27}$$

Moreover, using Proposition 12.2/5, we have

$$\alpha_{2k} = d_k'(0) = (2k+1)2^{2k} E_{2k}\left(\frac{1}{2}\right); \tag{14.28}$$

combining this with (8.104), we get

$$\alpha_{2k} = (2k+1)E_{2k}, \quad k = 0, 1, \dots, \tag{14.29}$$

where E_{2k} is the Euler number [113, p. 300]. In (14.29), α_{2k} is connected to the coefficient of the expansion of sech t, that is,

$$\sum_{k=0}^{\infty} \frac{\alpha_{2k}}{2k+1} \frac{t^{2k}}{(2k)!} = \operatorname{sech} t.$$

Hence, according to (12.6), we have

$$l(t) = \operatorname{sech} t \tag{14.30}$$

and

$$\frac{1}{l(t)} = \frac{1}{\operatorname{sech} t} = \cosh t = \sum_{k=0}^{\infty} \frac{t^{2k}}{(2k)!}. \tag{14.31}$$

But in (12.9), $\frac{1}{l(t)} = \sum_{k=0}^{\infty} \beta_{2k} \frac{t^{2k}}{(2k+1)!}$, and hence

$$\beta_{2k} = 2k + 1. \tag{14.32}$$

Now we get the generating function of the polynomial sequence $\{d_k\}_k$ defined in (12.7). As a result, we have

$$\frac{\operatorname{sech} t}{t} \sinh tx = \sum_{k=0}^{\infty} \frac{d_k(x)}{2k+1} \frac{t^{2k}}{(2k)!}. \tag{14.33}$$

For the conjugate sequence $\{\widehat{d}_k\}_k$, we have

$$t \cosh t \sinh tx = \sum_{k=0}^{\infty} \frac{\widehat{d}_k(x)}{2k+1} \frac{t^{2k}}{(2k)!}. \tag{14.34}$$

From (14.29) and (14.32), we have

$$d_k(x) = \sum_{i=0}^{k} \binom{2k+1}{2i+1} E_{2(k-i)} x^{2i+1},$$

$$\widehat{d}_k(x) = \sum_{i=0}^{k} \binom{2k+1}{2i+1} x^{2i+1}.$$

The conjugate sequence $\{\widehat{d}_k\}_k$ satisfies the conditions

$$\begin{cases} \widehat{d}_k(x) = (2k+1)(2k)\widehat{d}_{k-1}(x), \\ \widehat{d}_k(0) = 0, \quad \widehat{d}'_k(0) = 2k+1. \end{cases} \tag{14.35}$$

The polynomial sequence $\{d_k\}_k$ is not known in the literature, except [47, Example 3.3]; therefore we call it the *odd Lidstone–Euler polynomial sequence*, and we denote it with $\{\mathcal{E}_k\}_k$, whereas the conjugate sequence (also unknown) is denoted as $\{\widehat{\mathcal{E}}_k(x)\}_k$.

We give the first polynomials of these sequences:

$$\begin{aligned}
\mathcal{E}_0(x) &= x, \\
\mathcal{E}_1(x) &= -3x + x^3, \\
\mathcal{E}_2(x) &= 25x - 10x^3 + x^5, \\
\mathcal{E}_3(x) &= -427x + 175x^3 - 21x^5 + x^7, \\
\mathcal{E}_4(x) &= 12465x - 4284x^3 + 630x^5 - 36x^7 + x^9,
\end{aligned}$$

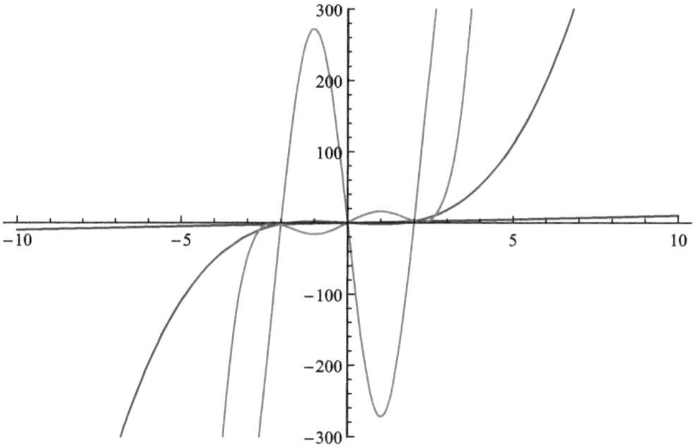

Figure 14.3: Lidstone–Euler polynomials.

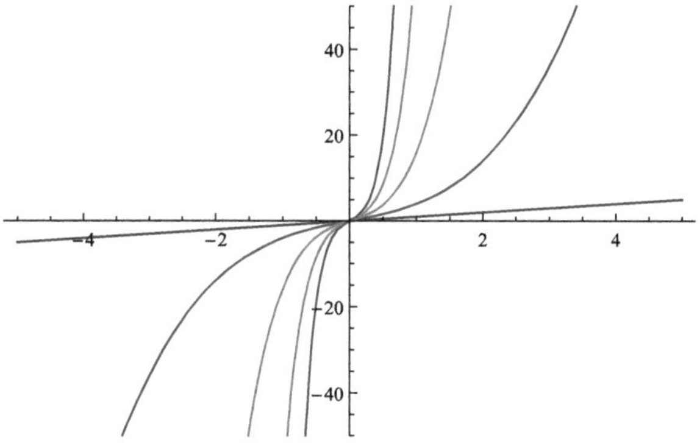

Figure 14.4: Lidstone–Euler conjugate polynomials.

$$\begin{aligned}
\widehat{\mathcal{E}}_0(x) &= x,\\
\widehat{\mathcal{E}}_1(x) &= 3x + x^3,\\
\widehat{\mathcal{E}}_2(x) &= 5x + 10x^3 + x^5,\\
\widehat{\mathcal{E}}_3(x) &= 7x + 35x^3 + 21x^5 + x^7,\\
\widehat{\mathcal{E}}_4(x) &= 9x + 84x^3 + 126x^5 + 84x^7 + x^9.
\end{aligned}$$

In Figures 14.3 and 14.4, we confront the graphs of these polynomials.

Considering (12.12) and (12.13), we have the recurrence relations

$$\mathcal{E}_k(x) = x^{2k+1} - \sum_{j=0}^{k-1} \binom{2k+1}{2j+1} \mathcal{E}_j(x), \tag{14.36}$$

$$\widehat{\mathcal{E}}_k(x) = x^{2k+1} - \sum_{j=0}^{k-1} \binom{2k+1}{2j+1} E_{2(k-j)} \widehat{\mathcal{E}}_j(x), \tag{14.37}$$

and the determinant forms

$$\mathcal{E}_k(x) = \frac{(-1)^k}{3!5!\cdots(2k-1)!}\begin{vmatrix} x & x^3 & x^5 & \cdots & x^{2k-1} & x^{2k+1} \\ 1 & 3 & 5 & \cdots & 2k-1 & 2k+1 \\ 0 & 3! & \frac{5!}{3!}3 & \cdots & \frac{(2k-1)!}{(2k-3)!}2k-3 & \frac{(2k+1)!}{(2k-1)!}2k-1 \\ \vdots & & \ddots & & \vdots & \vdots \\ \vdots & & & \ddots & \vdots & \vdots \\ 0 & \cdots & & \cdots & (2k-1)! & \frac{(2k+1)!}{3!}3 \end{vmatrix}, \tag{14.38}$$

$$\widehat{\mathcal{E}}_k(x) = \frac{(-1)^k}{3!5!\cdots(2k-1)!}$$

$$\times\begin{vmatrix} x & x^3 & x^5 & \cdots & x^{2k-1} & x^{2k+1} \\ E_0 & 3E_2 & 5E_4 & \cdots & (2k-1)E_{2(k-1)} & 2kE_{2k} \\ 0 & 3!E_0 & \frac{5!}{3!}3E_2 & \cdots & \frac{(2k-1)!}{(2k-3)!}(2k-3)E_{2(k-2)} & \frac{(2k+1)!}{(2k-1)!}(2k-1)E_{2(k-1)} \\ \vdots & & \ddots & & \vdots & \vdots \\ \vdots & & & \ddots & \vdots & \vdots \\ 0 & \cdots & & \cdots & (2k-1)!E_0 & \frac{(2k+1)!}{3!}3E_2 \end{vmatrix}. \tag{14.39}$$

To consider the odd Lidstone–Euler interpolation, we define the linear functional F on $C^{2k+2}[0,1]$ by

$$\mathrm{F}(f) = f'(1). \tag{14.40}$$

Then

$$\beta_{2i} = 2i + 1 = \mathrm{F}(x^{2i+1}). \tag{14.41}$$

Then using Theorem 13.2, we get the formula

$$f(x) = \mathcal{E}_k[f](x) + R_k[f](x), \tag{14.42}$$

where

$$\mathcal{E}_k[f](x) = \sum_{i=0}^{k} \frac{f^{(2i+1)}(1)}{(2i+1)!}\mathcal{E}_i(x), \tag{14.43}$$

and [56]

$$R_k[f](x) = -\sum_{i=0}^{k} \frac{f^{(2i)}(0)}{(2i+1)!}\mathcal{E}_i'(1-x) + \int_0^1 K(x,t)f^{(2k+2)}(t)dt, \tag{14.44}$$

where

$$K(x,t) = \frac{(x-t)_+^{2k+1}}{(2k+1)!} - \sum_{i=0}^{k} \frac{(1-t)^{2(k-i)+1}}{(2(k-i))!(2i+1)!} \varepsilon_i(x).$$

The following interpolant conditions are satisfied:

$$(\mathcal{E}_k[f](x))^{(2i+1)}(1) = f^{(2i+1)}(1), \quad i = 0, \dots, k. \tag{14.45}$$

We note that odd Lidstone–Euler polynomial interpolation (14.43) also satisfies homogeneous conditions in the origin, that is,

$$\mathcal{E}_k[f]^{(2i)}(0) = 0, \quad i = 0, \dots, k. \tag{14.46}$$

Then we can consider the polynomial

$$\overline{\mathcal{E}}_k[f](x) = f(0) + \sum_{i=0}^{k} \frac{f^{(2i+1)}(1)}{(2i+1)!} \mathcal{E}_i(x), \tag{14.47}$$

which satisfies the interpolatory conditions

$$\overline{\mathcal{E}}_k[f](0) = f(0), \quad (\overline{\mathcal{E}}_k[f])^{(2i+1)}(1) = f^{(2i+1)}(1), \quad i = 0, \dots, k. \tag{14.48}$$

The polynomial (14.47) can be used in the solution of the BVP:

$$\begin{cases} y^{(2k+2)}(x) = f(x, y(x), \dots, y^{(2k)}(x)), \\ y(0) = \alpha_0, \quad y^{(2i)}(0) = 0, \quad i = 1, \dots, k, \quad y^{(2i+1)}(1) = \beta_i, \quad i = 0, \dots, k. \end{cases} \tag{14.49}$$

It is possible to consider the interpolant polynomial

$$\hat{\varepsilon}_k[f](x) = \varepsilon_k[f](x) - \sum_{i=0} \frac{f^{(2i)}(0)}{(2i+1)!} \varepsilon_i'(1-x)$$

and to study the related interpolant properties. The details are in [56].

14.3 Second type or complementary Lidstone polynomials

Just as in the odd case, from Theorem 12.2, we consider the even Lidstone-type sequence $\{p_k\}_k$ defined by

$$p_0(x) = 1, \quad p_k(x) = 2^{2k} B_{2k}\left(\frac{x+1}{2}\right), \quad k = 1, \dots, \tag{14.50}$$

given the Bernoulli polynomials $B_k(x)$ of degree k.

By the well-known properties of the Bernoulli polynomials,

$$\begin{cases} p_k''(x) = (2k)(2k-1)p_{k-1}(x), & k = 1,\dots, \\ p_k'(0) = p_k'(1) = 0, & k > 1. \end{cases} \tag{14.51}$$

These polynomials are connected to the first type Lidstone polynomials. In fact, we get

$$p_k(x) = (2k)!\Lambda_k'(x), \quad k = 0,1,\dots. \tag{14.52}$$

Now, if

$$p_k(x) = \sum_{i=0}^{k} \binom{2k}{2i} \gamma_{2(k-i)} x^{2i}, \quad k = 0,1,\dots, \tag{14.53}$$

we have from (14.50)

$$\gamma_{2k} = p_k(0) = 2^{2k} B_{2k}\left(\frac{1}{2}\right) = 2^{2k}(2^{1-2k} - 1)B_{2k} = 2(1 - 2^{2k-1})B_{2k}. \tag{14.54}$$

Consequently, the generating function of the sequences (14.53) is $t \operatorname{csch} t \cosh tx$, that is,

$$t \operatorname{csch} t \cosh tx = \sum_{k=0}^{\infty} p_k(x) \frac{t^{2k}}{(2k)!}. \tag{14.55}$$

Moreover, if $\{\widehat{p}_k\}$ is the conjugate sequence of $\{p_k\}$, we get

$$\widehat{p}_k(x) = \sum_{i=0}^{k} \binom{2k}{2i} \delta_{2(k-i)} x^{2i}, \quad k = 0,1,\dots, \tag{14.56}$$

where $\{\gamma_{2k}\}$ and $\{\delta_{2k}\}$ satisfy the relation

$$\sum_{j=0}^{i} \binom{2i}{2j} \gamma_{2i} \delta_{2(i-j)} = \begin{cases} 1, & i = 0, \\ 0, & i > 0. \end{cases} \tag{14.57}$$

From (14.55), the generating function of $\{\widehat{p}_k\}$ is $\frac{\sinh t}{t} \cosh tx$, that is,

$$\frac{\sinh t}{t} \cosh tx = \sum_{k=0}^{\infty} \widehat{p}_k(x) \frac{t^{2k}}{(2k)!}. \tag{14.58}$$

The even polynomial sequence (14.53), up to a factorial constant $(2k)!$, is known in the literature as the *second type Lidstone polynomials* [68, 113, 128] or as *complementary Lidstone* [7, 48]. In both cases, it is denoted as $\{v_k\}_k$, that is, $v_k(x) = \frac{p_k(x)}{(2k)!}$.

The conjugate sequence is not known in the literature, and we denote it as $\{\widehat{v}_k\}_k$, that is, $\widehat{v}_k(x) = \frac{\widehat{p}_k(x)}{(2k)!}$.

We give the first polynomials of these sequences:

$$v_0(x) = 1,$$
$$v_1(x) = \frac{1}{2}x^2 - \frac{1}{6},$$
$$v_2(x) = \frac{1}{24}x^4 - \frac{1}{12}x^2 + \frac{7}{360},$$
$$v_3(x) = \frac{1}{720}x^6 - \frac{1}{144}x^4 + \frac{7}{720}x^2 - \frac{31}{15120},$$
$$v_4(x) = \frac{1}{40320}x^8 - \frac{1}{4320}x^6 + \frac{7}{8640}x^4 - \frac{31}{30240}x^2 + \frac{127}{604800},$$
$$v_5(x) = \frac{1}{3628800}x^{10} - \frac{1}{241920}x^8 + \frac{7}{259200}x^6 - \frac{31}{362880}x^4 + \frac{127}{1209600}x^2 - \frac{511}{23950080},$$

and

$$\widehat{v}_0(x) = 1,$$
$$\widehat{v}_1(x) = \frac{x^2}{2} + \frac{1}{6},$$
$$\widehat{v}_2(x) = \frac{x^4}{4!} + \frac{x^2}{2!3!} + \frac{1}{5!},$$
$$\widehat{v}_3(x) = \frac{x^6}{6!} + \frac{x^4}{4!3!} + \frac{x^2}{2!5!} + \frac{1}{7!},$$
$$\widehat{v}_4(x) = \frac{x^8}{8!} + \frac{x^6}{6!3!} + \frac{x^4}{4!5!} + \frac{x^2}{2!7!} + \frac{1}{9!},$$
$$\widehat{v}_5(x) = \frac{x^{10}}{10!} + \frac{x^8}{8!3!} + \frac{x^6}{6!5!} + \frac{x^4}{4!7!} + \frac{x^2}{2!9!} + \frac{1}{11!}.$$

In Figures 14.5 and 14.6, we show the graphs of these polynomials.

We note that $\{v_k\}_k$ and $\{\widehat{v}_k\}_k$ are not conjugate sequence; that is,

$$v_k(\widehat{v}_k(x)) \neq x^{2k}.$$

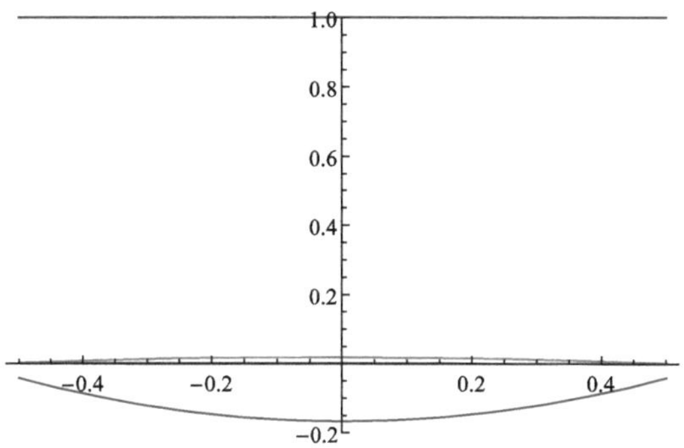

Figure 14.5: Even Lidstone polynomials.

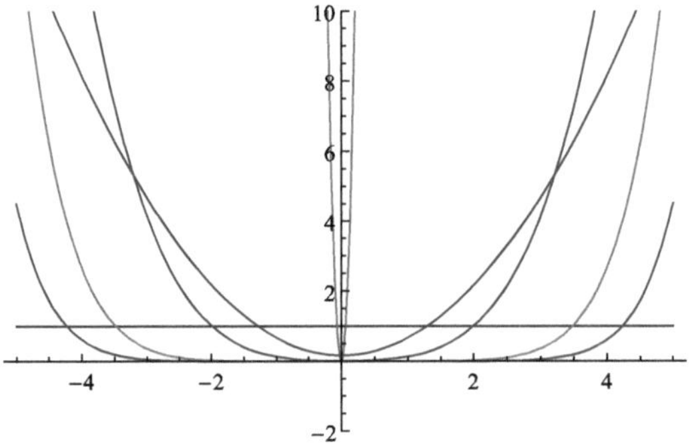

Figure 14.6: Even Lidstone conjugate polynomials.

For some properties of the sequence $\{\widehat{v}_k\}_k$, we refer to [48], because $\widehat{v}_k(x)$ satisfies the differential equation

$$\widehat{v}_k''(x) = \widehat{v}_{k-1}(x), \quad \widehat{v}_k'(0) = 0.$$

To consider the interpolant polynomials, let the linear functional F on $C^{2k+1}[0,1]$ be

$$F(f) = \int_0^1 f(t)dt; \tag{14.59}$$

then we have

$$\delta_{2i} = \frac{1}{(2i+1)} = F(x^{2i}) \quad \forall i = 0, 1, \dots. \tag{14.60}$$

The polynomial sequence related to the linear functional F is the sequence $\{v_k\}_k$, that is, the Lidstone II type sequence. Consequently, from (13.17), (13.18), (13.19), we get the following:

Theorem 14.1 ([48]). *For any $f \in C^{2k+1}[0,1]$, we have*

$$f(x) = P_0[f](x) + R_0[f](x), \tag{14.61}$$

where

$$P_0[f](x) = \int_0^1 f(t)dt + \sum_{i=1}^{k} (f^{(2i-1)}(1) - f^{(2i-1)}(0))v_i(x), \tag{14.62}$$

$$R_k[f](z) = \sum_{i=0}^{k-1} \frac{f^{(2i+1)}(0)}{(2i+1)!} R_k[x^{2i+1}](z) + \frac{1}{(2k+1)!} f^{(2k+1)}(\xi_z) R_k[x^{2k+1}](z), \tag{14.63}$$

and it is verified that

$$\begin{cases} \int_0^1 f(x)dx = \int_0^1 P_k[f](x)dx, \\ f^{(2i-1)}(1) - f^{(2i-1)}(0) = P_0^{2i-1}[f](1) - P_0^{2i-1}[f](0), \quad i = 1, \dots, k. \end{cases} \tag{14.64}$$

Remark 14.1. The expansion generated by (14.61) can be considered as the expansion of f in Bernoulli polynomials of even degree.

Applying Remark 12.8 and the above techniques, we get the following:

Theorem 14.2. *For any $f \in C^{2k+1}[0,1]$, we have*

$$f(x) = P_1[f](x) + R_1[f](x), \tag{14.65}$$

where

$$P_1[f](x) = \int_0^1 f(t)dt + \sum_{i=1}^{k} [f^{(2i-1)}(1)v_i(x) - f^{(2i-1)}(0)v_i(1-x)], \tag{14.66}$$

$$R_1[f](x) = f(x) - \int_0^1 f(t)dt + \sum_{i=1}^{k} [f^{(2i-1)}(1)v_i(x) - f^{(2i-1)}(0)v_i(1-x)]. \tag{14.67}$$

Moreover,

$$\begin{cases} \int_0^1 f(x)dx = \int_0^1 P_1[f](x)dx, \\ \frac{d^{2i-1}}{dx^{2i-1}} P_1[f](1) = f^{(2i-1)}(1), \quad i = 1, \dots, k, \\ \frac{d^{2i-1}}{dx^{2i-1}} P_1[f](0) = f^{(2i-1)}(0), \quad i = 1, \dots, k. \end{cases} \tag{14.68}$$

Finally, we have the interpolatory polynomials:

Theorem 14.3. *For any $f \in C^{2k+1}[0,1]$, we have*

$$f(x) = P_2[f](x) + R_2[f](x), \tag{14.69}$$

where

$$P_2[f](x) = f(0) + \sum_{j=1}^{k} [f^{(2j-1)}(1)(v_j(x) - v_j(0)) - f^{(2j-1)}(0)(v_j(1-x) - v_j(1))], \tag{14.70}$$

and it is verified that

$$\begin{cases} P_2[f](0) = f(0), \\ \frac{d^{(2i-1)} P_2[f](1)}{dx^{(2i-1)}} = f^{(2i-1)}(1), \quad i = 1, \dots, k, \\ \frac{d^{(2i-1)} P_2[f](0)}{dx^{(2i-1)}} = f^{(2i-1)}(0), \quad i = 1, \dots, k. \end{cases} \tag{14.71}$$

Proof. We easily verify that

$$P_2[f](x) = P_1[f](x) - f(0),$$

where $f(0)$ is given by (14.65). □

Remark 14.2. In Theorem 14.3, the role of the boundary values 0, 1 can be inverted, obtaining a symmetric formula with respect to (14.69), (14.70), (14.71).

Bounds of the remainder have been obtained. In fact:

Theorem 14.4 ([47]). *If* $f \in C^{2k+1}[0,1]$, *then for the linear functional* $R_2[f](x)$, *we have*

$$|R_2[f](x)| \leq \frac{k(2^{k+2}-1)}{(2k+2)!}|B_{2k+2}| \max_{0\leq x\leq 1}|f^{(2k+1)}(x)|. \tag{14.72}$$

Likewise, bounds of the remainder have been obtained in [209].

The expansion in series of Lidstone II-type polynomials can also be considered. In fact:

Theorem 14.5 ([68, p. 66]). *Any real entire function* $f(x)$ *of exponential type less than* π *has the absolutely and uniformly convergent expansion*

$$f(x) = \int_0^1 f(x)dx + \sum_{n=1}^{\infty}[f^{(2n-1)}(1)v_n(x) - f^{(2n-1)}(0)v_n(1-x)]. \tag{14.73}$$

Lidstone II-type interpolation polynomials have been applied to numerical summation and quadrature [7, 54] and to even high-order differential equation with boundary conditions (14.71).

14.4 Even Lidstone–Euler polynomials

By symmetry to odd Lidstone–Euler polynomials, we consider the even polynomial sequence $\{p_k\}$ defined by

$$p_k(x) = 2^{2k}E_{2k}\left(\frac{x+1}{2}\right), \quad k = 0,1\ldots, \tag{14.74}$$

where $E_k(x)$ is the Euler polynomial of degree k.

By the well-known proprieties of Euler polynomials,

$$\begin{cases} p_k''(x) = (2k)(2k-1)p_{k-1}(x), & k = 1,\ldots, \\ p_k(1) = p_k'(0) = 0, & k = 1,\ldots. \end{cases} \tag{14.75}$$

This sequence is connected to the odd Lidstone–Euler polynomials by

$$p_k(x) = \frac{\mathcal{E}_k'(x)}{2k+1}, \quad k = 0,\ldots. \tag{14.76}$$

For the sequence $\{p_k\}_k$, the generating function is $\operatorname{sech} t \cosh tx$, that is,

$$\operatorname{sech} t \cosh tx = \sum_{k=0}^{\infty} p_k(x)\frac{t^{2k}}{2k!}. \tag{14.77}$$

In fact, setting

$$h(t) = \sum_{k=0}^{\infty} \gamma_{2k} \frac{t^{2k}}{(2k)!}$$

with

$$\gamma_{2k} = p_k(0) = 2^{2k} E_{2k}\left(\frac{1}{2}\right) = E_{2k},$$

given E_{2k}, the Euler number, we get

$$h(t) = \sum_{k=0}^{\infty} \gamma_{2k} \frac{t^{2k}}{(2k)!} = \sum_{k=0}^{\infty} \frac{E_{2k}}{(2k)!} t^{2k} = \operatorname{sech} t. \tag{14.78}$$

For the conjugate polynomial sequence $\{\widehat{p}_k\}_k$, we get the generating function

$$\cosh t \cosh tx = \sum_{k=0}^{\infty} \widehat{p}_k(x) \frac{t^{2k}}{(2k)!}. \tag{14.79}$$

In fact,

$$\frac{1}{h(t),} = \cosh t = \sum_{k=0}^{\infty} \frac{t^{2k}}{(2k)!} = \sum_{k=0}^{\infty} \delta_{2k} \frac{t^{2k}}{(2k)!},$$

with

$$\delta_{2k} = 1.$$

We call the polynomial sequence $\{p_k\}$ the *even Lidstone–Euler polynomial sequence*, and we denote it as $\{\mathcal{S}_k\}_k$, whereas we denote the conjugate sequence as $\{\widehat{\mathcal{S}}_k\}_k$. We note that the sequence $\{\mathcal{S}_k\}_k$ coincide (less than a constant) with the polynomials $\mu_i(x)$, $i = 0, 1, \dots,$ of the modified Abel series [149].

The first polynomials of these sequences are as follows:

$$\begin{aligned}
&\mathcal{S}_0(x) = 1,\\
&\mathcal{S}_1(x) = x^2 - 1,\\
&\mathcal{S}_2(x) = x^4 - 6x^2 + 5,\\
&\mathcal{S}_3(x) = x^6 - 15x^4 + 75x^2 - 61,\\
&\mathcal{S}_4(x) = x^8 - 28x^6 + 350x^4 - 1708x^2 + 1385,\\
&\mathcal{S}_5(x) = x^{10} - 45x^8 + 1050x^6 + 12810x^4 + 62325x^2 - 50521,
\end{aligned}$$

and

$$\begin{aligned}
\widehat{\mathcal{S}}_0(x) &= 1,\\
\widehat{\mathcal{S}}_1(x) &= x^2 + 1,\\
\widehat{\mathcal{S}}_2(x) &= x^4 + 6x^2 + 1,\\
\widehat{\mathcal{S}}_3(x) &= x^6 + 15x^4 + 15x^2 + 1,\\
\widehat{\mathcal{S}}_4(x) &= x^8 + 28x^6 + 70x^4 + 28x^2 + 1,\\
\widehat{\mathcal{S}}_5(x) &= x^{10} + 45x^8 + 210x^6 + 210x^4 + 45x^2 + 1.
\end{aligned}$$

In Figures 14.7 and 14.8, we compare the graphs of these polynomials.

Remark 14.3. From (14.79), we get

$$\widehat{\mathcal{S}}_k(x) = \frac{(1+x)^{2k} + (1-x)^{2k}}{2} \quad \forall k \in \mathbb{N}. \tag{14.80}$$

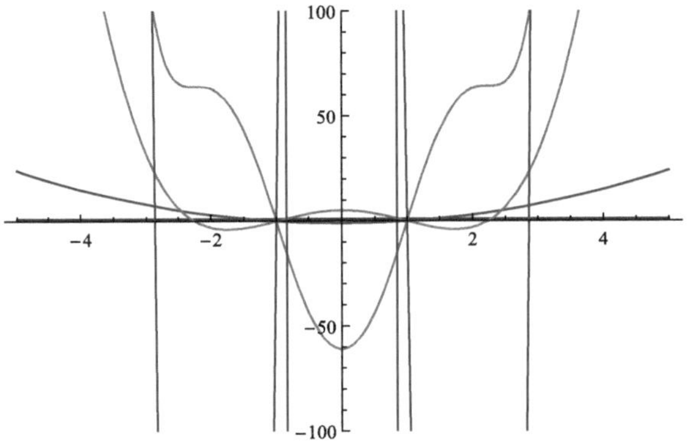

Figure 14.7: Lidstone–Euler polynomials.

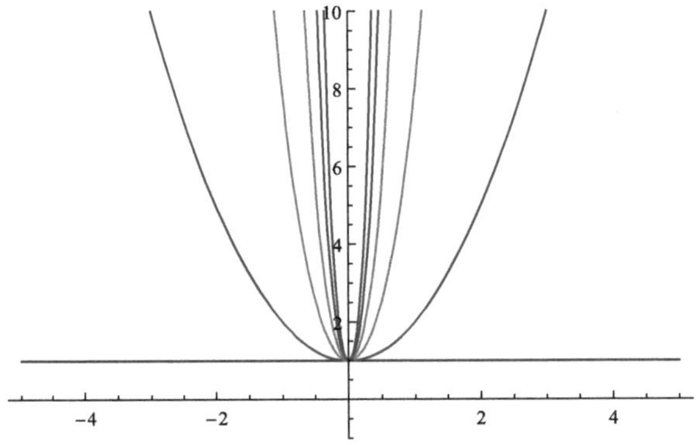

Figure 14.8: Lidstone–Euler conjugate polynomials.

We note that the sequence $\{\widehat{\mathcal{S}}_k\}_k$ satisfies the BVP

$$\begin{cases} \widehat{\mathcal{S}}_k''(x) = (2k)(2k-1)\widehat{\mathcal{S}}_{k-1}(x), \\ \widehat{\mathcal{S}}_k(0) = 1, \quad \widehat{\mathcal{S}}_k'(0) = 0. \end{cases} \tag{14.81}$$

Now we obtain the recurrence relations

$$\mathcal{S}_k(x) = x^{2k} - \sum_{j=0}^{k-1} \binom{2k}{2j} \mathcal{S}_j(x), \tag{14.82}$$

$$\widehat{\mathcal{S}}_k(x) = x^{2k} - \sum_{j=0}^{k-1} \binom{2k}{2j} E_{2(k-j)} \widehat{\mathcal{S}}_j(x), \tag{14.83}$$

and the determinantal form

$$\mathcal{S}_0(x) = 1,$$

$$\mathcal{S}(x) = (-1)^k \begin{vmatrix} 1 & x^2 & x^4 & \cdots & x^{2k-2} & x^{2k} \\ 1 & 1 & 1 & \cdots & 1 & 1 \\ 0 & 1 & \binom{4}{2} & \cdots & \binom{2k-2}{2} & \binom{2k}{2} \\ \vdots & & \ddots & & \vdots & \vdots \\ \vdots & & & \ddots & \vdots & \vdots \\ 0 & \cdots & & \cdots & 1 & \binom{2k}{2(k-1)} \end{vmatrix}, \tag{14.84}$$

and

$$\widehat{\mathcal{S}}_0(x) = 1$$

$$\widehat{\mathcal{S}}_k(x) = (-1)^k \begin{vmatrix} 1 & x^2 & x^4 & \cdots & x^{2k-2} & x^{2k} \\ E_0 & E_2 & E_4 & \cdots & E_{2k-2} & E_{2k} \\ 0 & E_0 & \binom{4}{2}E_2 & \cdots & \binom{2k-2}{2}E_{2(k-2)} & \binom{2k}{2}E_{2(k-1)} \\ \vdots & & \ddots & & \vdots & \vdots \\ \vdots & & & \ddots & \vdots & \vdots \\ 0 & \cdots & & \cdots & E_0 & \binom{2k}{2(k-1)}E_2 \end{vmatrix}. \tag{14.85}$$

We note that from (14.83), by taking $x = 0$, we get the remarkable recursive formula for the Euler numbers:

$$E_{2k} = -1 - \sum_{j=1}^{k-1} \binom{2k}{2j} E_{2(k-j)}, \quad k = 2, 3, \ldots. \tag{14.86}$$

From the determinantal form (14.85), we get the interesting identities

$$(-1)^k \begin{vmatrix} 1 & x^2 & x^4 & \cdots & x^{2k-2} & x^{2k} \\ E_0 & E_2 & E_4 & \cdots & E_{2k-2} & E_{2k} \\ 0 & E_0 & \binom{4}{2}E_2 & \cdots & \binom{2k-2}{2}E_{2(k-2)} & \binom{2k}{2}E_{2(k-1)} \\ \vdots & & \ddots & & \vdots & \vdots \\ \vdots & & & \ddots & \vdots & \vdots \\ 0 & \cdots & & \cdots & E_0 & \binom{2k}{2(k-1)}E_2 \end{vmatrix} = \frac{(1+x)^{2k}+(1-x)^{2k}}{2} \tag{14.87}$$

and

$$(-1)^k \begin{vmatrix} E_2 & E_4 & \cdots & E_{2k-2} & E_{2k} \\ E_0 & \binom{4}{2}E_2 & \cdots & \binom{2k-2}{2}E_{2(k-2)} & \binom{2k}{2}E_{2(k-1)} \\ \vdots & \ddots & & \vdots & \vdots \\ \vdots & & \ddots & \vdots & \vdots \\ 0 & \cdots & \cdots & E_0 & \binom{2k}{2(k-1)}E_2 \end{vmatrix} = 1. \tag{14.88}$$

To consider the interpolation polynomial, let F be the linear functional on $C^{2k}[0,1]$ defined by

$$F(f) = f(1). \tag{14.89}$$

Then assuming that

$$\delta_{2i} = F(x^{2i}) = 1, \tag{14.90}$$

the polynomials related to F are $\mathcal{S}_i(x)$. The interpolant polynomial is

$$\mathcal{S}_k[f](x) = \sum_{i=0}^{k} \frac{f^{(2i)}(1)}{(2i)!}\mathcal{S}_i(x) = f(1) + \sum_{i=1}^{k} \frac{f^{(2i)}(1)}{(2i)!}\mathcal{S}_i(x), \tag{14.91}$$

which satisfies

$$\mathcal{S}_k^{(2i)}[f](1) = f^{(2i)}(1), \quad i = 0,\dots,k. \tag{14.92}$$

If we now consider

$$\overline{\mathcal{S}}_k[f](x) = \sum_{i=0}^{k} \frac{f^{(2i)}(0)}{(2i)!}\mathcal{S}_i(1-x), \tag{14.93}$$

we get

$$\overline{\mathcal{S}}_k^{(2i)}[f](0) = f^{(2i)}(0), \quad i = 0,\dots,k. \tag{14.94}$$

Now we get the following:

Theorem 14.6. *The polynomial*

$$\widehat{\widehat{\mathcal{S}}}_k[f](x) = f(0) + \sum_{i=0}^{k} \frac{f^{(2i)}(1)}{(2i)!}(\mathcal{S}_i(x) - \mathcal{S}_i(0)) \tag{14.95}$$

satisfies the interpolant conditions

$$\widehat{\widehat{\mathcal{S}}}_k[f](0) = f(0), \quad \widehat{\widehat{\mathcal{S}}}_k[f]^{(2i)}(1) = f^{(2i)}(1), \quad i = 0, \dots, k. \tag{14.96}$$

Proof. It follows from (14.91) and (14.92). □

A formula analogous to (14.95), with respect to point 1 is possible; then we have the following formula:

$$\widehat{\widehat{\mathcal{S}}}_k[f](x) = f(1) + \sum_{i=0}^{k} \frac{f^{(2i)}(0)}{(2i)!}(\mathcal{S}_i(x) - \mathcal{S}_i(1)). \tag{14.97}$$

As already mentioned in the context of operators approximation theory involving Sheffer polynomials, the sequence $\{\widehat{\mathcal{S}}_k\}_k$ has been considered in [41]. The convergence of the related Favard–Szasz operators has been proved, but it is possible to apply the extrapolation process to accelerate the rate of convergence. Some numerical examples will be presented in a separate paper.

Bibliography

[1] Abbasbandy S., Shivanian E., A new analytical technique to solve Fredholm's integral equations, Numer. Algorithms, **56**, 27–43 (2011).

[2] Abel U., Ivan M., Asymptotic expansion of the Jakimovski–Leviatan operators and their derivatives, In: L. Leindler, F. Schipp, J. Szabados (eds.) Functions, series, operators, 103–119, Budapest (2002).

[3] Aceto L., Cacão I., A matrix approach to Sheffer polynomials, J. Math. Anal. Appl., **446**, 87–100 (2017).

[4] Aceto L., Malonek H. R., Tomaz G., A unified matrix approach to the representation of Appell polynomials, Integral Transforms Spec. Funct., **26**, 426–441 (2015).

[5] Agapito J., Mestre A., Petrullo L., Torress M., A symbolic treatment of Riordan array, Linear Algebra Appl., **439**, 1700–1715 (2013).

[6] Agarwal R. P., O'Regan D., Lidstone continuous and discrete boundary-value problems, Mem. Differ. Equ. Math. Phys., **19**, 107–125 (2000).

[7] Agarwal R. P., Pinelas S., Wong P. J. Y., Complementary Lidstone interpolation and boundary value problems, J. Inequal. Appl. (2009), doi:10.1155/2009/624631.

[8] Agarwal R. P., Wong P. J. Y., Lidstone polynomials and boundary value problems, Comput. Math. Appl., **17**, 1397–1421 (1989).

[9] Agarwal R. P., Wong P. J. Y., Error inequalities in polynomial interpolation and their applications, Springer science & business media, **262** (2012).

[10] Agratini O., Binomial polynomials and their applications in approximation theory. Conf. Semin. Mat. Univ. Bari, **281**, 1–22 (2001).

[11] Aitken A. C., A general formula of polynomial interpolation. Proc. Edinb. Math. Soc., **1(4)**, 199–203 (1929).

[12] Al-Salam W. A., q-Appell polynomials, Ann. Mat. Pura Appl., **77**, 31–45 (1967).

[13] Altomare F., Campiti M., Korovkin-type approximation theory and its applications, De Gruyter stud. math., **17** (1994).

[14] Altomare M., Costabile F. A., A new determinant form of Bessel polynomials and apllications. Math. Comput. Simul., **141**, 16–23 (2017).

[15] Alzer H., Sharp bounds for the Bernoulli numbers, Arch. Math. (Basel) **74**, 207–211 (2000).

[16] Andrews G. E., Uchimura K., Identities in combinatorics IV: differentiation and harmonic numbers, Util. Math., **28**, 265–269 (1985).

[17] Appell P., Sur une classe de polynômes. Ann. Sci. Éc. Norm. Supér., 119–144 (1880).

[18] Avram F., Taqqu M. S., Noncentral limit theorems and Appell polynomials, Ann. Probab., 767–775, (1987).

[19] Barry P., On a family of generalized Pascal triangles defined by exponential Riordan arrays, J. Integer Seq., **10(3)** (2007).

[20] Bateman H., Higher transcendental functions. The Bateman manuscript project, Vol. I–III, McGraw-Hill, New York (1953).

[21] Bell E. T., Invariant sequences, Proc. Natl. Acad. Sci. USA, **14(12)**, 901–904 (1928).

[22] Bell E. T., Exponential polynomials, Ann. Math., **35**, 258–277, (1934).

[23] Bell E. T., The history of Blissard's symbolic method, with a sketch of its inventor's life, Am. Math. Mon., **45**, 414–421 (1938).

[24] Bernoulli J., Ars Conjectandi, Opus Posthumum; Accedit Tractatus De Seriebus Infinitis, Et Epistola Gallicè scripta De Ludo Pilae Reticularis. Thurnisii (1713).

[25] Bernstein S., Démonstration du théorème de Weierstrass fondée sur le calcul des probabilités, Comm. Soc. Math. Kharkow, **13**, 1–2 (1912).

https://doi.org/10.1515/9783110652925-015

[26] Bhrawy A. H., Tohidi E., Soleymani F., A new Bernoulli matrix method for solving high-order linear and nonlinear Fredholm integro-differential equations with piecewise intervals, Appl. Math. Comput., **219**, 482–497 (2012).

[27] Boas Jr R. P., Representation of functions by Lidstone series, Duke Math. J., **10**, 239–245 (1943).

[28] Boas R. P., Buck R. C., Polynomial expansions of analytic functions, Springer (2013).

[29] Boyadzhiev K. N., Exponential polynomials, Stirling numbers, and evaluation of some gamma integrals, Abstr. Appl. Anal. (2009), doi:10.1155/2009/168672.

[30] Bretti G., Ricci P. E., Multidimensional extensions of the Bernoulli and Appell polynomials, Taiwan. J. Math., **8**, 415–428 (2004).

[31] Buckholtz J. D., Shaw J. K., On functions expandable in Lidstone series, J. Math. Anal. Appl., **47**, 626–632 (1974).

[32] Canfield E. R., Central and local limit theorems for the coefficients of polynomials of binomial type, J. Comb. Theory, **23**, 275–290 (1977).

[33] Carlitz L., A note on the Bessel polynomials, Duke Math. J., **24**, 151–162 (1957).

[34] Carlitz L., Some polynomials of Touchard connected with the Bernoulli numbers, Can. J. Math., **9**, 188–190 (1957).

[35] Chebyshev P. L., Sur le développement des fonctions à une seule variable, Bull. Acad. Sci. St. Petersb., 193–200 (1859).

[36] Cheney E. W., Sharma A., On a generalization of Bernstein polynomials, Riv. Mat. Univ. Parma, **5**, 77–84 (1964).

[37] Cheney E. W., Sharma A., Bernstein power series, Can. J. Math. **16**, 241–252 (1964).

[38] Cheng S. X., Characterization for binomial sequences among renewal sequences, Appl. Math.-JCU, **7(1)**, 114–128 (1992).

[39] Cheon G. S., Kim H., Shapiro L. W.: Riordan group involutions, Linear Algebra Appl., **428**, 941–952 (2008).

[40] Chihara T. S., An introduction to orthogonal polynomials, Dover Publications, New York (1978).

[41] Ciupa A., Approximation by a generalized Szász type operator, J. Comput. Anal. Appl., **5**, 413–424 (2003).

[42] Comtet L., Advanced combinatorics: the art of finite and infinite expansions, Springer science & business media (2012).

[43] Costabile F., Expansions of real functions in Bernoulli polynomials and applications, Conf. sem., University Bari, **273** (1999).

[44] Costabile F., Dell'Accio F., Gualtieri M. I., A new approach to Bernoulli polynomials, Rend. Mat. Appl., **26**, 1–12 (2006).

[45] Costabile F., Gualtieri M. I., Serra S., Asymptotic expansion and extrapolation for Bernstein polynomials with applications, BIT, **36**, 676–687 (1996).

[46] Costabile F., Gualtieri M. I., Serra S., Asymptotic expansion and extrapolation for positive operators, Ann. Un. Ferrara, Sez. VII - Sc. Math., **XLV**, 431–442 (2000).

[47] Costabile F., Napoli A., A special class of polynomials related to non-classic general interpolatory problems, Integral Transforms Spec. Funct., **20**, 539–550 (2009).

[48] Costabile F., Napoli A., Special even polynomials and related interpolatory problems, Integral Transforms Spec. Funct., **21**, 183–196 (2010).

[49] Costabile F., Napoli A., A class of collocation methods for numerical integration of initial value problems, Comput. Math. Appl., **62**, 3221–3235 (2011).

[50] Costabile F., Napoli A., Collocation for high-order differential equations with lidstone boundary conditions, J. Appl. Math. (2012), doi:10.1155/2012/120792.

[51] Costabile F. A., Un teorema di separazione degli zeri dei polinomi ultrasferici, Boll. Unione Mat. Ital., **12–A** (1976).

[52] Costabile F. A., Lezioni di Istituzioni di Matematiche, mod. I, Liguori, Napoli (1999).

[53] Costabile F. A., Dell'Accio F., Expansion over a rectangle of real functions in Bernoulli polynomials and applications, BIT, **41**, 451–464 (2001).

[54] Costabile F. A., Dell'Accio F., Luceri R., Explicit polynomial expansions of regular real functions by means of even order Bernoulli polynomials and boundary values, J. Comput. Appl. Math., **176**, 77–90 (2005).

[55] Costabile F. A., Gualtieri M. I., Napoli A., Some results on generalized Szász operators involving Sheffer polynomials, J. Comput. Appl. Math., **337**, 244–255 (2018).

[56] Costabile F. A., Gualtieri M. I., Napoli A., Odd and even Lidstone-type polynomial sequences. Part 2: applications, submitted.

[57] Costabile F. A., Gualtieri M. I., Napoli A., Altomare M., Odd and even Lidstone-type polynomial sequences. Part 1: basic topics, Adv. Differ. Equ. (2018), doi:10.1186/s13662-018-1733-5.

[58] Costabile F. A., Gualtieri M. I., Serafini G., Cubic Lidstone-spline for numerical solution of BVPs, Math. Comput. Simul., **141**, 56–64 (2017), doi:10.1016/j.matcom.2017.001.006.

[59] Costabile F. A., Gualtieri M. I., Serra Capizzano S., An iterative method for the computation of the solutions of nonlinear equations, Calcolo, **36**, 17–34 (1999).

[60] Costabile F. A., Longo E., A determinantal approach to Appell polynomials, J. Comput. Appl. Math., **234**, 1528–1542 (2010).

[61] Costabile F. A., Longo E., The Appell interpolation problem, J. Comput. Appl. Math., **236**, 1024–1032 (2011).

[62] Costabile F. A., Longo E., Algebraic theory of Appell polynomials with application to general linear interpolation problem, In: A. Yasser (ed.) Linear algebra, 21–46, InTech Europe Publication, Rijeka, Croatia (2012).

[63] Costabile F. A., Longo E., Δ_h-Appell sequences and related interpolation problem, Numer. Algorithms, **63**, 165–186 (2013).

[64] Costabile F. A., Longo E., An algebraic exposition of umbral calculus with application to general linear interpolation problem—a survey, Publ. Inst. Math., Nouv. Sér., **96(110)**, 67–83 (2014).

[65] Costabile F. A., Longo E., An algebraic approach to Sheffer polynomial sequences, Integral Transforms Spec. Funct., **25(4)**, 295–311 (2014).

[66] Costabile F. A., Longo E., A new recurrence relation and related determinantal form for binomial type polynomial sequences, Mediterr. J. Math., **13(6)**, 4001–4017 (2016).

[67] Costabile F. A., Longo E., Umbral interpolation, Publ. Inst. Math., Nouv. Sér, Beograd, **99**, 165–175 (2016).

[68] Costabile F. A., Longo E., Luceri R., A new proof of uniform convergence of Bernoulli and Lidstone series for entire real functions of exponential type, Rendiconti dell'Istituto Lombardo Accademia di Scienze e Lettere, Classe di Scienze Matematiche Fisiche e Naturali, Settore: Matematica, Milano, **143**, 63–70 (2009).

[69] Costabile F. A., Napoli A., A new spectral method for a class of linear boundary value problems, J. Comput. Appl. Math. **292**, 329–341 (2016).

[70] Costabile F. A., Serpe A., An algebraic approach to Lidstone polynomials, Appl. Math. Lett., **20**, 387–390 (2007).

[71] Costabile F. A., Serpe A., Bruzio A., No classic boundary conditions, In: Proc. WCE (2007), London, UK, July 2–4 (2007).

[72] Craciun M., Approximation operators constructed by means of Sheffer sequences, Rev. Anal. Numér. Théor. Approx., **30**, 135–150 (2001).

[73] Craciun M., On compound operators constructed with binomial and Sheffer sequences, Rev. Anal. Numér. Théor. Approx., **32(2)**, 135–144 (2003).
[74] Craciun M., Di Bucchianico A., Sheffer sequences, probability distributions and approximation operators, SPOR Report **4** (2005).
[75] Davis H. T., The theory of linear operators from the standpoint of differential equations of infinite order, Bloomington, Indiana (1936).
[76] Davis P. J., Interpolation and approximation, Dover (1975).
[77] Deutsch E., Ferrari L., Rinaldi S., Production matrices and Riordan arrays, Ann. Comb., **13**, 65–85 (2009).
[78] Di Bucchianico A., Probabilistic and analytical aspects of the umbral calculus, CWI Tracts, **119**, 1–148 (1997).
[79] Di Bucchianico A., Loeb D., A selected survey of umbral calculus, Electron. J. Comb., **2** (2000).
[80] Di Bucchianico A., Loeb D. E., Sequences of binomial type with persistent roots, J. Math. Anal. Appl., **199**, 39–58 (1996).
[81] Di Nardo E., Niederhausen H., Senato D., A symbolic handling of Sheffer polynomials, Ann. Mat. Pura Appl., **190(3)**, 489–506 (2011).
[82] Di Nardo E., Senato D., Umbral nature of the Poisson random variables, In: Algebraic combinatorics and computer science, Springer, 245–266 (2001).
[83] Di Nardo E., Senato D., An umbral setting for cumulants and fractional moments, Eur. J. Comb., **27(3)**, 394–413 (2006).
[84] Dilcher K., Asymptotic behaviour of Bernoulli, Euler, and generalized Bernoulli polynomials, J. Approx. Theory, **49**, 321–330 (1987).
[85] Djordjevic G. B., Milovanovic G. V., Special classes of polynomials, Univerzitet u Niu, Tehnoloki fakultet, Leskovac (2014).
[86] Dolgy D. V., Kim D. S., Kim T., Korobov polynomials of the first kind (Russian), Math. Sb., **208(1)**, 65–79 (2017), traslation in Sb. Math., **208(1–2)**, 60–74 (2017).
[87] Dong Quan N., The classical umbral calculus and the flow of a Drinfeld module, Trans. Am. Math. Soc., **369(2)**, 1265–1289 (2017).
[88] Elezović N., Generalized Bernoulli polynomials and numbers, revisited, Mediterr. J. Math., **13**, 141–151 (2016).
[89] Euler L., Methodus generalis summandi progressiones, Comment. Acad. Sci. Imp. Petrop., **6** (1738).
[90] Favard J., Sur les polynomes de Tchebichef, C. R. Acad. Sci., Paris, **200**, 2052–2053 (1935).
[91] Fillmore J. P., Williamson S. G., A linear algebra setting for the Rota–Mullin theory of polynomials of binomial type, Linear Multilinear Algebra, **1**, 67–80 (1973).
[92] Galiffa D. J., On the higher-order Sheffer orthogonal polynomial sequences, Springer science & business media (2013).
[93] Garsia A. M., An exposé of the Mullin–Rota theory of polynomials of binomial type, Linear Multilinear Algebra, **1**, 47–65 (1973).
[94] Ge H. F., New sharp bounds for the Bernoulli numbers and refinement of Becker–Stark inequalities, J. Appl. Math. (2012), doi:10.1155/2012/137507.
[95] Gershgorin S. A., Uber die Abgrenzung der Eigenwerte einer matrix, Izv. Ahad. Navk., **6**, 749–754 (1931).
[96] Gessel I. M., Applications of the classical umbral calculus, Algebra Univers. (2003), doi:10.1007/s00012-003-1813-5.
[97] Golightly G. O., Absolutely convergent Lidstone series, J. Math. Anal. Appl., **125**, 72–80 (1987).
[98] Golightly G. O., A note on a series of Lidstone polynomials, J. Math. Anal. Appl., **158**, 556–573 (1991).

[99] Gould H. W., Stirling number representation problems, Proc. Am. Math. Soc., **11**, 447–451 (1960).
[100] Gould H. W., Combinatorial identities: a standardized set of tables listing 500 binomial coefficient summations, Morgantown, W Va (1972).
[101] Harper L. H., Stirling behavior is asymptotically normal, Ann. Math. Stat., **38(2)**, 410–414 (1967).
[102] He M. X., Ricci P. E., Differential equation of Appell polynomials via the factorization method, J. Comput. Appl. Math., **139**, 231–237 (2002).
[103] He T. X., Hsu L. C., Shiue P. J. S., The Sheffer group and the Riordan group, Discrete Appl. Math., **155**, 1895–1909 (2007).
[104] He T. X., Sprugnoli R., Sequence characterization of Riordan arrays, Discrete Math., **309**, 3962–3974 (2009).
[105] Higham N. J., Accuracy and stability of numerical algorithms, SIAM (2002).
[106] Howard F., Roots of the Euler polynomials, Pac. J. Math., **64**, 181–191 (1976).
[107] Inkeri K., The real roots of Bernoulli polynomials, Ann. Univ. Turku. Ser. AI, **37**, 1–20 (1959).
[108] Isaacson E., Keller H. B., Analysis of numerical methods, John Wiley and Sons, Inc., New York (1966).
[109] Islam M. D., Shirin A. et al., Numerical solutions of a class of second order boundary value problems on using Bernoulli polynomials, arXiv preprint arXiv:1309.6064 (2013).
[110] Ismail M. E. H., On a generalization of Szász operators, Mathematica (Cluj), **39**, 259–267 (1974).
[111] Ismail M. E. H., Polynomials of binomial type and approximation theory, J. Approx. Theory, **23**, 177–186 (1978).
[112] Jakimovski A., Leviatan D., Generalized Szász operators for the approximation in the infinite interval, Mathematica (Cluj), **34**, 97–103 (1969).
[113] Jordan K., Calculus of finite differences, Amer. Math. Soc. (1965).
[114] Kac V., Cheung P., Quantum calculus, Springer science & business media (2001).
[115] Katz V. J., A history of mathematics, Pearson/Addison-Wesley (2004).
[116] Khan S., Raza N., 2-Iterated Appell polynomials and related numbers, Appl. Math. Comput., **219**, 9469–9483 (2013).
[117] Khan S., Riyasat M., A determinantal approach to Sheffer–Appell polynomials via monomiality principle, J. Math. Anal. Appl., **421**, 806–829 (2015).
[118] Khan S., Riyasat M., Determinantal approach to certain mixed special polynomials related to Gould–Hopper polynomials, Appl. Math. Comput., **251**, 599–614 (2015).
[119] Kim D. S., Kim T., An umbral calculus approach to Bernoulli–Padé polynomials, In: Mathematical analysis, approximation theory and their applications, Springer optim. appl., **11**, 363–382, Springer, Cham (2016).
[120] Kim S. D., Kim T., A matrix approach to some identities involving Sheffer polynomial sequences, Appl. Math. Comput., **253**, 83–101 (2015).
[121] Kim T., Kim S. D., Identities for degenerate Bernoulli polynomials and Korobov polynomials of the first kind, Sci. China Math. (2018), doi:10.1007/s11425-018-9338-5.
[122] Laplace P. S., Mémoire sur les approximations des formules qui sont fonctions de très-grands nombres, et sur leur application aux probabilités, Baudouin (1810).
[123] Leeming D. J., The real zeros of the Bernoulli polynomials, J. Approx. Theory, **58**, 124–150 (1989).
[124] Lehmer D. H., On the maxima and minima of Bernoulli polynomials, Am. Math. Mon., **47**, 533–538 (1940).
[125] Lehmerm D. H., A new approach to Bernoulli polynomials, Am. Math. Mon., **95**, 905–911 (1988).

[126] Levi D., Tempesta P., Winternitz P., Lorentz and Galilei invariance on lattices, Phys. Rev. D, **69(10)**, 105011 (2004).

[127] Levi D., Tempesta P., Winternitz P., Umbral calculus, difference equations and the discrete Schrödinger equation, J. Math. Phys., **45**, 4077–4105 (2004).

[128] Lidstone G. J., Notes on the extension of Aitken's theorem (for polynomial interpolation) to the Everett types. Proc. Edinb. Math. Soc., **2(2)**, 16–19 (1930).

[129] Lorentz G. G., Bernstein polynomials, American Mathematical Soc. (2012).

[130] Loureiro A. F., Maroni P., Around q-Appell polynomial sequences, Ramanujan J., **26**, 311–321 (2011).

[131] Lucas E., Theorie des nombres, Vol. 1, Gauthier-Villars (1891).

[132] Luke Y. L., Special functions and their approximations, Vol. 2, Academic Press, New York (1969).

[133] Lupas A., Approximation operators of binomial type, In: New developments in approximation theory, 175–198, Springer, Birkhäuser, Basel (1999).

[134] Lupas L., Lupas A., Polynomials of Binomial type and approximation operators, Studia Univ. Babes-Bolyai Math. XXXII, **4**, 61–63 (1987).

[135] Markowsky G., Differential operators and the theory of binomial enumeration, J. Math. Anal. Appl., **63**, 145–155 (1978).

[136] Maroni P., Mejri M., Generalized Bernoulli polynomials revisited and some other Appell sequences, Georgian Math. J., **12**, 697–716 (2005).

[137] Merlini D., Rogers D. G., Sprugnoli R., Verri M. C., On some alternative characterizations of Riordan arrays, Can. J. Math., **49**, 301–320 (1997).

[138] Merlini D., Sprungoli R., Verri M. C., The Cauchy numbers, Discrete Math., **306**, 1906–1920 (2006).

[139] Mihoubi M., Bell polynomials and binomial type sequences, Discrete Math., **308**, 2450–2459 (2008).

[140] Monroe J. L., Extrapolation and the Bulirsch–Stoer algorithm, Phys. Rev. E., **65** (2002), doi:10.1103/PhysRevE.65.066116.

[141] Mühlbach G., Operatoren vom Bernsteinschen Type, J. Approx. Theory, **3**, 274–292 (1970).

[142] Mullin R., Rota G. C., On the foundations of combinatorial theory: theory of binomial enumeration, In: B. Harrie (ed.) Graph theory and its applications, 167–213, Academic Press, New York–London (1970).

[143] Mursaleen M., Ansari K. J., Approximation by generalized Szász operators involving Sheffer polynomials, arXiv preprint, arXiv:1601.00675 (2015).

[144] Natalini P., Bernardini A., A generalization of the Bernoulli polynomials, J. Appl. Math., **3**, 155–163 (2003).

[145] Nemes G., An asymptotic expansion for the Bernoulli numbers of the second kind, J. Integer Seq., **14**, Article 11.4.8 (2011).

[146] Niederhausen H., Finite operator calculus with applications to linear recursions, Florida Atlantic University, Boca Raton (2010).

[147] Nkwanta A., A Riordan matrix approach to unifying a selected class of combinatorial arrays, Congr. Numer., 33–46 (2003).

[148] Pan V., Structured matrices and polynomials: unified superfast algorithms, Springer Sci. (2012).

[149] Pethe S. P., Sharma A., Modified Abel expansion and a subclass of completely convex functions, J. Math. Anal., **3**, 546–558 (1972).

[150] Picone M., Fichera G., Corso di Analisi Matematica I. Veschi Ed., Roma (1972).

[151] Polya G., Über die funktionalgleichung der exponentialfunktion im matrixkalkul, Sitzber. Preuss. Akad. Wiss, 96–99 (1928).

[152] Popa E. C., Sheffer polynomials and approximation operators, Tamkang J. Math., **34**, 117–128 (2003).
[153] Popa E. C., On a sequence of approximation operators constructed by means of generating functions for binomial polynomials, In: Proceedings of the 11th WSEAS international conference on computers, Agios Nikolas, Creta Island (Greece), July 26–28 (2007).
[154] Popoviciu T., Remarques sur les polynômes binomiaux, Bull. Soc. Math. Cluj, **6**, 8–10 (1932).
[155] Portisky H., On certain polynomial and other approximations to analytic functions, Proc. Natl. Acad. Sci. USA, **16**, 83–85 (1930).
[156] Qi F., Explicit formulas for computing Bernoulli numbers of the second kind and Stirling numbers of the first kind, Filomat **28(2)**, 319–327 (2014), doi:10.2298/FIL1402319O.
[157] Qi F., An integral representation and properties of Bernoulli numbers of the second kind, Bull. Korean Math. Soc., **52**, 987–998 (2015).
[158] Qi F., A new formula for the Bernoulli numbers of the second kind in terms of the Stirling numbers of the first kind, Publ. Inst. Math. Nouv. Sér., (Beograd), **100(104)**, 243–249 (2016).
[159] Qi F., Chapman R. J., Two closed forms for the Bernoulli polynomials, J. Number Theory, **159**, 89–100 (2016).
[160] Qi F., Zhao J. L., Some properties of the Bernoulli numbers of the second kind and their generating function, J. Differ. Equ. Appl. (2016).
[161] Raabe J. L., Zurückführung einiger Summen und bestimmten Integrale auf die Jacob-Bernoullische Function, J. Reine Angew. Math., **42**, 348–367 (1851).
[162] Rad J. A., Kazem S., Shaban M., Parand K., A new operational matrix based on Bernoulli polynomials, arXiv preprint arXiv:1408.2207 (2014).
[163] Rainville E. D., Special functions, Vol. 8, The Macmillan Co., New York (1960).
[164] Riordan J., Combinatorial identities, Wiley (1968).
[165] Roman S., The theory of the umbral calculus II, J. Math. Anal. Appl., **89**, 290–314 (1982).
[166] Roman S., The umbral calculus, Academic Press, New York (1984).
[167] Roman S. M., Rota G. C., The umbral calculus, Adv. in Math., **27**, 95–188 (1978).
[168] Rota G. C., Kahaner D., Odhyzko A., On the foundations of Combinatorial theory VII. Finite operator calculus, J. Math. Anal. Appl., **42**, 684–760 (1973).
[169] Rota G. C., Shen J., On the combinatorics of cumulants, J. Comb. Theory, Ser. A, **91(1–2)**, 283–304 (2000).
[170] Rota G. C., Taylor B. D., The classical umbral calculus, SIAM J. Math. Anal., **25(2)**, 694–711 (1994).
[171] Saadatmandi A., Dehghan M., A new operational matrix for solving fractional-order differential equations, Comput. Math. Appl., **59**, 1326–1336 (2010).
[172] Sablonniere P., Positive Bernstein–Sheffer operators, J. Approx. Theory, **83**, 330–341 (1995).
[173] Schoenberg I. J., On certain two-point expansions of integral functions of exponential type, Bull. Am. Math. Soc., **42**, 284–288 (1936).
[174] Schoenberg I. J., Contributions to the problem of approximation of equidistant data by analytic functions. Part A. On the problem of smoothing or graduation. A first class of analytic approximation formulae, Q. Appl. Math., **4**, 45–99 (1946).
[175] Schoenberg I. J., Contributions to the problem of approximation of equidistant data by analytic functions. Part B. On the problem of osculatory interpolation. A second class of analytic approximation formulae, Q. Appl. Math., **4**, 112–141 (1946).
[176] Schur I., On Faber polynomials, Am. J. Math., **67(1)**, 33–41 (1945).
[177] Shapiro L. W., Getu S., Woan W. J., Woodson L., The Riordan group, Discrete Appl. Math., **34**, 229–239 (1991).
[178] Sharma A., Chak A. M., The basic analogue of a class of polynomials, Riv. Mat. Univ. Parma, **5**, 325–337 (1954).

[179] Sheffer I. M., Some properties of polynomial sets of type zero, Duke Math. J., **5**, 590–622 (1939).

[180] Shohat J., The relation of the classical orthogonal polynomials to the polynomials of Appell, Am. J. Math., **58**, 453–464 (1936).

[181] Srivastava H. M., Todorov P. G., An explicit formula for the generalized Bernoulli polynomials, J. Math. Anal. Appl., **130**, 509–513 (1988).

[182] Stancu D. D., Approximation of functions by a new class of linear polynomial operators, Rev. Roum. Math. Pures Appl., **13**, 1173–1194 (1968).

[183] Stancu D. D., Approximation of functions by means of some new classes of positive linear operators, In: Numerische Methoden der Approximationstheorie, 187–203, Springer, Birkhäuser, Basel (1972).

[184] Steffens K. G., The history of approximation theory: from Euler to Bernstein, Springer science business media (2007).

[185] Steffensen J. F., The poweroid, an extension of the mathematical notion of power, Acta Math., **73**, 333–366 (1941).

[186] Steffensen J. F., Interpolation, Courier Corporation (2006).

[187] Stetter H. J., Analysis of discretitation methods for differential equation, Springer, Berlin (1973).

[188] Sucu S., Büyükyazici J., Integral operators containing Sheffer polynomials, Bull. Math. Anal. Appl., **4**, 56–66 (2012).

[189] Sucu S., İbikli E., Rate of convergence for Szász type operators including Sheffer polynomials, Stud. Univ. Babeş–Bolyai, Math., **58**, 55–63 (2013).

[190] Szász O., Generalization of S. Bernstein's polynomials to the infinite interval, J. Res. Natl. Bur. Stand., **45**, 239–245 (1950).

[191] Szëgo G., Orthogonal polynomials, Amer. Math. Soc. (1939).

[192] Ta B. Q., Probabilistic approach to Appell polynomials, Expo. Math., **33**, 269–294 (2015).

[193] Taylor B. D., Umbral presentation for polynomial sequences, Comput. Math. Appl., **41(9)**, 1085–1089 (2001), ISSN0898-1221. Umbral Calculus and its applications (Cortona 1998).

[194] Tempesta P., On Appell sequences of polynomials of Bernoulli and Euler type, J. Math. Anal. Appl., **341**, 1295–1310 (2008).

[195] Tohidi E., Kılıçman A., A collocation method based on the Bernoulli operational matrix for solving nonlinear BVPs which arise from the problems in calculus of variation, Math. Probl. Eng. (2013), doi:10.1155/2013/757206.

[196] Varma R. S., On Appell polynomials, Proc. Am. Math. Soc., **2**, 593–596 (1951).

[197] Vein P. R., Matrices which generate families of polynomials and associated infinite series, J. Math. Anal. Appl., **59**, 278–287 (1977).

[198] Vein P. R., Identities amog certain triangular matrices, Linear Algebra Appl., **82**, 27–79 (1986).

[199] Verde-Star L., Inverses of generalized Vandermonde matrices, J. Math. Anal. Appl., **131**, 341–353 (1988).

[200] Verde-Star L., Polynomial sequences of interpolatory type, Stud. Appl. Math., **53**, 153–171 (1993).

[201] Verde-Star L., Infinite lower triangular matrices, q-Pascal matrices, and determinantal representations, Linear Algebra Appl., **434**, 307–318 (2011).

[202] Voronovskaya E., Détermination de la forme asymptotique de l'approximation des fonctions par les polynômes de M. Bernstein, C. R. (Dokl.) Acad. Sci. URSS, **4**, 86–92 (1932).

[203] Walz G., Asymptotic and extrapolation, Akademic Verlag, Berlin (1996).

[204] Wazwaz A. M., A reliable treatment of singular Emden–Fowler initial value problems and boundary value problems, Appl. Math. Comput., **217**, 10387–10395 (2011).

[205] Whittaker J. M., On Lidstone's series and two-point expansions of analytic functions, Proc. Lond. Math. Soc., **2(1)**, 451–469 (1934).
[206] Whittaker J. M., Interpolatory function theory, Cambridge tracts in math. physics, **33** (1936).
[207] Widder D. V., Functions whose even derivatives have a prescribed sign, Proc. Natl. Acad. Sci. USA, **26**, 657–659 (1940).
[208] Widder D. V., Completely convex functions and Lidstone series, Trans. Am. Math. Soc., **51**, 387–398 (1942).
[209] Wong P. J. Y., On Lidstone splines and some of their applications, Neural Parallel Sci. Comput., **1**, 472–475 (1995).
[210] Wong P. J. Y., Agarwal R. P., Sharp error bounds for the derivatives of Lidstone-spline interpolation, Comput. Math. Appl., **28**, 23–53 (1994).
[211] Wong P. J. Y., Agarwal R. P., Sharp error bounds for the derivatives of Lidstone-spline interpolation II, Comput. Math. Appl., **31**, 61–90 (1996).
[212] Yang S., Some identities involving the binomial sequences, Discrete Math., **308**, 51–58 (2008).
[213] Yang S. L., Recurrence relations for the Sheffer sequences, Linear Algebra Appl., **437**, 2986–2996 (2012).
[214] Yang Y., Youn H., Appell polynomial sequences: a linear algebra approach, JP J. Algebra Number Theory Appl., **13(1)**, 65–98, (2009).
[215] Youn H., Yang Y., Differential equation and recursive formulas of Sheffer polynomial sequences, ISRN Discrete Math (2011), doi:10.5402/2011/476462.
[216] Yousefi S. A., Dehghan M., The use of He's variational iteration method for solving variational problems, Int. J. Comput. Math., **87**, 1299–1314 (2010).
[217] Zatorskii R. A., Malyarchuk A. R., Triangular matrices and combinatorial inversion formulas, Math. Notes, **85**, 11–19 (2009).

Index

Abel
- δ-operator 60
- binomial identity 60
- conjugate polynomial sequence 57
- determinantal form 58
- generating function 61
- interpolation 61
- polynomial sequence 57
- recurrence relation 59

Appell
- approximation operators 103
- conjugate polynomial sequence 79
- creation matrix 95
- derivation matrix 95
- differential equation 87
- first determinantal form 85
- first recurrence relation 83
- generalized Appell identity 90
- generating function 80
- integration matrix 95
- interpolation 101
- polynomial sequence 78
- second determinantal form 86
- second recurrence relation 84

Appell–Laguerre
- conjugate polynomial sequence 141
- determinantal form 143
- generating function 142
- interpolation 144
- polynomial sequence 141
- recurrence relation 143

Bell polynomials 24

Bernoulli
- approximation operators 121
- conjugate numbers 109
- conjugate polynomial sequence 108
- derivation matrix 110
- determinantal form 113
- generalization 122
- generating function 112
- integration matrix 111
- interpolation 120
- multiplication theorem 122
- numbers 109
- polynomial sequence 108
- recurrence relation 112

Bernoulli of second kind
- determinantal form 177
- generating function 182
- polynomial sequence 179
- recurrence relation 177

Bernstein operator 42

Bessel polynomial 45

Bessel reverse polynomial 45

binomial
- conjugate polynomial sequence 25
- derivation matrix 29
- differential equation 24
- first determinantal form 31
- first recurrence relation 28
- generating function 26
- interpolation 41
- operators approximation 42
- polynomial sequence 24
- second determinantal form 32
- second recurrence relation 29
- third recurrence relation 31

binomial Laguerre
- binomial identity 65
- determinantal form 64
- generating function 66
- operator approximation 67
- polynomials 64
- recurrence relation 64

binomial Laguerre,
- δ-operator 66

Boole
- conjugate polynomial sequence 193
- generating function 193
- interpolation 194
- polynomial sequence 192
- recurrence relation 192

Cauchy numbers of first type 180

central factorial
- δ-operator 70
- determinantal form 72
- interpolation 71
- polynomials 70
- recurrence relation 72

Charlier polynomials 156

https://doi.org/10.1515/9783110652925-016

conjugate sequence 8
convolution product 79
Cramer's rule 32

δ-functional 34
δ-operator 35
determinantal form 192

Euler
– approximation operators 133
– binomial identity 131
– conjugate polynomial sequence 124
– determinantal form 129
– generating function 127
– integration matrix 126
– interpolation 133
– numbers 128
– polynomial sequence 124
– recurrence relation 127
even Lidstone
– approximation operators 221
– conjugate polynomial sequence 210
– determinantal form 213
– generating function 212
– interpolation 219
– polynomials 210
– recurrence relation 213
– relation with Appell polynomial 213
exponential
– δ-operator 54
– binomial identity 54
– determinantal form 52
– generating function 52
– operators sequence 55
– polynomial sequence 52
– recurrence relation 53
exponential-type operator 44

falling polynomials 51
Favard, theorem 139
Favard–Szasz operator 220
formal power series 7

Gaussian elimination 33
generalized Laguerre polynomials 63
generating function 8

harmonic numbers 110
Hermite
– approximation operators 140
– conjugate polynomial sequence 135
– determinantal form 138
– generating function 137
– integration matrix 136
– interpolation 140
– polynomial sequence 135
– recurrence relation 138
Hessemberg determinant 114
Hessemberg matrix 19

inverse, compositional 9
inverse, matrix 14
invertible series 19

Krawtchouk polynomials 156

Laguerre
– determinantal form 187
– interpolation 189
– polynomial sequence 186
– recurrence relation 187
lower factorial
– δ-operator 54
– binomial identity 54
– determinantal form 52
– generating function 52
– polynomial sequence 51
– recurrence relation 53

matrix
– Appell-type 10
– binomial-type 7
– conjugate 8
– even Lidstone-type 17
– exponential Riordan 13
– odd Lidstone-type 15
– production 18
– Sheffer-type 12
– Toeplitz 11
mean, operator 131
Meixener polynomials 156
Meixener–Pollaczek polynomials 156

Newton polynomial 55

odd Lidstone
– conjugate polynomial sequence 204
– determinantal form 208
– generating function 205
– interpolation 217
– polynomials 204

– recurrence relation 207
– relation with Appell 207
orthogonal polynomials 31

Poisson–Charlier
– conjugate polynomial sequence 197
– differential equation 197
– generating function 198
– interpolation 198
– polynomial sequence 197
– recurrence relation 197

q-Appell 98
q-calculus 98
q-derivative 98

Riordan group 20

Sheffer
– approximation operators 172
– conjugate polynomial sequence 150
– derivation 158
– differential equation 161
– differentiation matrix 162
– first determinantal form 154
– first recurrence relation 154
– generating function 153
– integration 161
– polynomial sequence 150
– second determinantal form 156
– second recurrence relation 154
– Taylor–Sheffer polynomials 168
– third determinantal form 157
– third recurrence relation 157
Stirling number
– generating function 52
– of second kind 51

umbral composition 39

W-Lambert function 57

De Gruyter Studies in Mathematics

Volume 34/1
József Lőrinczi, Fumio Hiroshima, Volker Betz
Feynman-Kac-Type Theorems and Gibbs Measures: Volume 1
ISBN 978-3-11-033004-5, e-ISBN 978-3-11-033039-7, e-ISBN (ePUB) 978-3-11-038993-7

Volume 34/2
József Lőrinczi, Fumio Hiroshima
Applications in Rigorous Quantum Field Theory: Volume 2
ISBN 978-3-11-040350-3, e-ISBN 978-3-11-040354-1, e-ISBN (ePUB) 978-3-11-040360-2

Volume 71
Wolfgang Herfort, Karl H. Hofmann, Francesco G. Russo
Periodic Locally Compact Groups: A Study of a Class of Totally Disconnected Topological Groups
ISBN 978-3-11-059847-6, e-ISBN 978-3-11-059919-0, e-ISBN (ePUB) 978-3-11-059908-4

Volume 70
Christian Remling
Spectral Theory of Canonical Systems
ISBN 978-3-11-056202-6, e-ISBN 978-3-11-056323-8, e-ISBN (ePUB) 978-3-11-056228-6

Volume 69
Derek K. Thomas, Nikola Tuneski, Allu Vasudevarao
Univalent Functions. A Primer
ISBN 978-3-11-056009-1, e-ISBN 978-3-11-056096-1, e-ISBN (ePUB) 978-3-11-056012-1

Volume 68/1
Timofey V. Rodionov, Valeriy K. Zakharov
Set, Functions, Measures. Volume 1: Fundamentals of Set and Number Theory
ISBN 978-3-11-055008-5, e-ISBN 978-3-11-055094-8

Volume 68/2
Alexander V. Mikhalev, Timofey V. Rodionov, Valeriy K. Zakharov
Set, Functions, Measures. Volume 2: Fundamentals of Functions and Measure Theory
ISBN 978-3-11-055009-2, e-ISBN 978-3-11-055096-2

Volume 67
Alexei Kulik
Ergodic Behavior of Markov Processes. With Applications to Limit Theorems
ISBN 978-3-11-045870-1, e-ISBN 978-3-11-045893-0

www.degruyter.com

Bei Fragen zur Produktsicherheit wenden Sie sich bitte an:
If you have any questions regarding product safety,
please contact:

Walter de Gruyter GmbH
Genthiner Straße 13
10785 Berlin
productsafety@degruyterbrill.com